Bicycle Transit
Its Planning and Design

Bruce L. Balshone
Paul L. Deering
Brian D. McCarl

foreword by
Tom McCall

The Praeger Special Studies program—utilizing the most modern and efficient book production techniques and a selective worldwide distribution network—makes available to the academic, government, and business communities significant, timely research in U.S. and international economic, social, and political development.

Bicycle Transit
Its Planning and Design

PRAEGER SPECIAL STUDIES • DESIGN/ENVIRONMENTAL PLANNING SERIES

Praeger Publishers   New York   Washington   London

Library of Congress Cataloging in Publication Data

Balshone, Bruce L
    Bicycle transit, its planning and design.

    (Praeger special studies, design/environmental
planning series)
    Bibliography: p.
    Includes index.
    1. Cycling paths—Design and construction.   I. Deering,
Paul L. , joint author.   II. McCarl, Brian D. , joint
author.   III. Title.
TE301. B34   711'. 7   75-55
ISBN 0-275-05410-1

PRAEGER PUBLISHERS
111 Fourth Avenue, New York, N.Y. 10003, U.S.A.

Published in the United States of America in 1975
by Praeger Publishers, Inc.

Printed in the United States of America

This is a book whose time came when Oregon passed the nation's first law setting aside highway-user revenue for construction of bicycle trails and footpaths. That law provided needed recognition of a pressing national need.

Bicycle Transit: Its Planning and Design is a most comprehensive work and is meticulous in detail. It makes a strong case for the bicycle as a primary means of transportation.

The private automobile is not on its last legs as the predominant means of personal transportation. But because of the auto's dependence on nonrenewable resources it is essential that we adopt more conservation-oriented methods of moving people. Our answers appear to lie chiefly in the past: buses, railroads, and bicycles. This book explains why and how we should reaccommodate the bicycle in transit planning.

The authors' intention is to provide a catalyst for discussion and citizen involvement in the planning and design process. I believe this book can have that impact.

Tom McCall

In the fall of 1973, a research team within the Urban Planning and Landscape Architecture Departments of the University of Oregon joined together as the Oregon Bicycle Transit Study Committee under my leadership, with the assistance of Paul L. Deering and Brian D. McCarl. The committee's goals were to evaluate and assimilate the bicycle studies of various governmental planning agencies, universities, and private consultants, and then produce a text which could be used by professionals and students as a guideline for establishing bicycle planning and design criteria. It is hoped that this book will serve as a comprehensive resource for the investigation of bicycle systems history, planning, engineering, legislation, and design concepts.

Bruce L. Balshone

Few people can write a comprehensive reference book without major assistance from persons representing numerous professional disciplines. In addition to thanking these generous people, the authors would like to express special gratitude to still others who gave their time and knowledge to help prepare this manuscript for publication:

A note of gratitude is due to Ting Li Cho, Associate Professor of Urban Planning, University of Oregon, and Jerome Diethelm, Associate Professor of Landscape Architecture, University of Oregon; both contributed planning and design recommendations.

To Roberta Deering, Department of Urban Planning, University of Oregon, who served as a research assistant in the preliminary writing of this manuscript.

To Donald P. Cato, Department of Landscape Architecture, University of Oregon, who helped provide technical and visual continuity.

To Vickie Golden, who edited the manuscript for style, and James M. Longstreth, who provided the illustrations.

To George Russell Norris, who translated bibliographical references, thus aiding research. To David Wolf and Janet Baer, who reviewed foreign bibliographical references and further contributed to the research process.

To Wallace M. Ruff, Professor of Landscape Architecture, University of Oregon, who served as a horticultural advisor in reviewing the plant material recommendations.

To Karen C. Cato, for her endless patience and cooperation in typing and editing the numerous preliminary drafts.

Page

FOREWORD
Tom McCall

PREFACE
Bruce L. Balshone

ACKNOWLEDGMENTS

LIST OF TABLES

LIST OF FIGURES

INTRODUCTION

Chapter

LIST OF TABLES

LIST OF FIGURES

Cities throughout the United States and abroad have shown growing concern over the creation of comprehensive transit systems—systems flexible enough to accommodate new modes of transportation as well as older forms like the increasingly popular bicycle. Dramatic growth patterns and changes within urban and rural communities have made it imperative that we institute responsive land-use planning; unless measures are taken now to insure highly coordinated and integrated transit systems, we will find it very difficult to make alterations in our transportation networks to suit our changing needs. In connection with this goal of comprehensive transit planning, the authors have brought together many current ideas and projects born of recent bikeway and urban planning studies, both in this country and abroad. This book includes investigations into the history, planning, engineering, legislation, and design concepts of bikeway systems, as well as an extensive research bibliography.

Design and planning are not endeavors solely for experts with credentials in those fields. The future of all planning and design depends upon participation from those people with the most accurate and carefully considered information: those who live within and use the system. Thus the planning and design of comprehensive transit systems must be responsive to the individual needs of bicyclists, motorists, and all people actively involved within the total transportation network.

It is true that state and local planners are becoming more and more aware of the need to combine the knowledge of landscape architects, planners, engineers, and sociologists when formulating new transit systems or modifying old ones. For instance, highways have a definite impact on urban social and environmental patterns; the physical means and avenues through which people mobilize themselves determine their opportunities for social involvement and contact. So highway planners are responsible for developing systems which directly affect the quality and substance of people's lives and must call upon a broad range of knowledge in order to create successful systems. However, while planning and design groups have lately organized within a richly diversified frame of professional reference, unfortunately they have remained isolated and centralized within the state or local hierarchy. With this isolation, planners and designers have cut off input from the people who live, work, and travel through the environments which are subject to their planning and design. These professional groups' commitment to a particular environment's needs and their awareness of its character can never be as great as that of the people who actually use it.

In order to maintain the crucially important communication that people have with their environment, they must be allowed to control any change which takes place there. As mentioned, an environment's character profoundly affects the development of people's lives. In turn, people profoundly affect the character and development of their environment. When the process of exchange between people and environment is interrupted or taken over by a body outside of the neighborhood, such as a local planning agency, the resulting change tends not to build upon the environment's previous historical experience or spatial arrangement, or upon the intrinsic elements which combined to create that place. Confronted by this kind of change, people withdraw from their environment and leave themselves lost in a disconnected, unresponsive maze.

Rather than allow a neighborhood's traditions and unique elements to become disassociated from the people who live there, planning and design must spring from within the neighborhood. Change which is to be valid in the historical perspective of past and future must build upon present conditions, allowing people's past experience to help create their environment. The process of design and planning should arise out of a succession of sequential acts, with each helping further to clarify the character of the place in which design and planning occur. If planning and design of comprehensive bicycle transportation systems begin at the neighborhood level, the greatest opportunity for truly coordinated and environmentally responsive systems is possible.

The essential quality of bicycle transportation planning and design which is begun at the neighborhood level will be retained when connections are made with other neighborhood, local, intercity, and regional areas. The conscious involvement of people within a specific neighborhood will be manifested in their system and the interconnections of their system with adjoining neighborhoods. As a result, the total local and intercity network will reflect the same concern for carefully considered detail as that shown at the neighborhood level.

Planning and design, which respond to a neighborhood's socio-economic needs and to the special requirements of its physical environment and structure, logically begin with bicycles and pedestrians. The bicycle is the vehicular mode most suitable and consistent with the development of transportation within a neighborhood—for example, transit to school and recreation areas. It is also ideal for travel from the neighborhood to outlying recreation centers, places of employment, and shopping centers. The main purpose of this book, then, is to provide a comprehensive resource which can evoke the participation of the bicycle user. Rather than formulate inflexible master plans for particular bikeway systems, the intention has been to provide a catalyst for creative expression and for satisfaction of public needs and desires. Involving the public in the design process will result in more viable transit systems.

Bicycle Transit
Its Planning and Design

EARLY HISTORY—THE FIRST BICYCLES

The current worldwide interest in bicycles is not a new phenomenon. Since the origin of the machine, there have been several peaks and depressions in the bicycle movement. It is speculated that the first bicycle may have been developed in China hundreds of years ago. However, there is also evidence that the Egyptians used a primitive, two-wheel vehicle, and many consider this to be the predecessor of the modern bicycle. French records of an Egyptian obelisk describe the hieroglyphic figure of a man seated on a horizontal bar mounted on two wheels. This obelisk, it is claimed, dates back to the reign of Rameses II, between the 14th and 13th centuries B.C.[1]

In England, historians have located evidence of 17th-century bicycle use. A stained glass window in the churchyard at Stoke Poges, four miles south of Windsor, dates from the 1600s and portrays a male figure straddling a vehicle with a saddle connecting two large wheels.[2]

In the late 18th century in France, the Count of Surac created his own version of a bicycle. It was a vehicle with two wheels joined by a wood frame resembling the body of a horse. Parisians were first introduced to Surac's vehicle in 1790, when he rode out from an alley of chestnut trees at the Jardin du Palais Royal atop his strange two-wheel mount, pushing himself along with his feet.[3]

FIGURE 1

Seventeenth-Century Stained Glass Window

Seventeenth-century stained glass window, Stoke Poges, Great Britain.

FIGURE 2

French Hobby Horse

French hobby horse, Paris, France, 1790.

It was not until 1818, when a French pioneer photographer named Nicéphore Niepce developed the *celerifère* (fast-feet), that any further innovation in the bicycle occurred. Niepce's vehicle, which consisted of two wheels fastened together by a beam, with a cross bar used as a handrest, was first exhibited in Luxembourg Gardens in the spring of 1818.[4]

At approximately the same time Niepce was working in France, eastward along the Rhine River the Baron Karl von Drais de Saverbrun was trying to figure out how to make the *celerifère* easier to propel. He added "a fork for the front wheel, which permitted the machine to be steered by the handlebars."[5] Patent laws early in the 19th century were weak, and many French as well as English copies of von Drais's machine could be found. A London coachman by the name of Dennis Johnson added an adjustable saddle and armrest to minimize the strain on the forearms,[6] and in 1855 an iron farrier by the name of Pierre Michaux created what we now call pedals and pedal-shafts.[7]

Soon after, another Frenchman, Pierre Michaux's associate Pierre Lallement, designed the *pont-a-mousson,* better known as the "boneshaker"—a device which the American historian R. A. Smith describes in his book *The Social History of the Bicycle:* "It con-

FIGURE 3

Celerifère (Fast-Feet)

Celerifère (fast-feet), developed by Nicéphore Niepce, Paris, France, 1818.

sisted of two iron-tired wooden wheels mounted one behind the other, a front fork and handlebar to permit steering, pedals on the axle of the front wheel, and a saddle fastened to a wooden frame with steel springs."[8] In 1866, Lallement immigrated to Ansonia, Connecticut, where he formed a partnership with James Carrol and began production of the "boneshaker." The next year he returned to France to exhibit his new model at the Universal Exposition of Paris, and while there he formed a new partnership with his former boss Pierre Michaux.[9]

The experimentation continued. One of the most dramatic alterations in bicycle design was the high-wheeler, called the "ordinary." In May, 1869, the English firm of Reynolds and May showed the Phanton model at the Crystal Palace in London. Another innovation two years later was the introduction of a rubber

FIGURE 4

Pont-a-mousson (the "Boneshaker")

Pont-a-mousson (the "boneshaker"), developed by Pierre Lallement, Paris, France, 1867.

cover over the metal pedals, so that the bicyclist could use the ball of his foot rather than the instep. Steel rims were also developed at this time, which enabled Parisian designers to increase the size of the front wheel and reduce the diameter of the rear wheel. This change was announced when a new model by John Keen was displayed at the Philadelphia Summer Centennial Exposition in 1876.[10]

FIGURE 5

High-wheeler (the "Ordinary")

High-wheeler (the "ordinary"), Great Britain, 1869.

Albert A. Pope, a former Civil War officer from Boston, viewed the bicycle at this exposition and was so impressed that he made a special trip to Britain to tour the various bicycle manufacturing plants. Upon his return to the United States in 1877, he converted the Pope Manufacturing Company into a bicycle import house and began to manufacture the Columbia model ordinary, which weighed 70 pounds and sold for $313.[11]

In an attempt to gain public acceptance of the device, Charles E. Pratt, one of the earliest bicyclists and an admirer of Pope, wrote the first handbook for bicyclists, called *The American Bicycler.* Prior to this time, bicyclists simply had no rights on the nation's roadways.[12] Also, to fill the needs and protect the rights of bicyclists, Pope, Pratt, and other avid supporters of the bicycle movement began to form bicycle clubs on the East Coast. Later, parallel organizations were established throughout the nation. The most prominent national club was the League of American Wheelmen, which originated in 1880 in Newport, Rhode Island. The League's major aims were to seek rights for the bicyclist and to improve the nation's roadways.

THE FORMATION OF BICYCLE CLUBS

The bicycle was both a symbol and vehicle for etiquette, and to some extent proper moral conduct, in 19th-century America. American churchmen of the 1890s considered it far better recreation than that to be found in the poolrooms of the day, and in the opinions of many the bicycle could unite entire families in an atmosphere of outdoor recreation. Above all, the bicycle could bring together fathers and sons and help to bridge the generation gap.

Americans and their European counterparts rallied around the bicycle and banded together in a great assortment of clubs. These organizations were a manifestation of the growing camaraderie among bicyclists of the day. While they did draw a great deal of attention and did result in the proliferation of festive apparel, these organizations were not merely groups of people seeking identification through fancy uniforms. Cycling clubs were formed for the explicit purpose of active growth and for the advocacy and promotion of the bicycle as a means of transportation.

The first national bicycle conference, organized and sponsored by the various East Coast bicycle clubs, was held in Newport, Rhode Island, on May 31, 1880. The conference precipitated formation of the national organization, the League of American Wheelmen (L.A.W.). At its inception, the L.A.W. pledged to repeal any and all laws that might interfere with the development of facilities for bicycle touring.

At the second annual L.A.W. conference, held in Boston in 1881, Lewis J. Bates of Michigan urged the League to take an active role in support of the national movement for better roads. Prior to this time, no federal agency existed to build or maintain a national roadway system. The roads which did exist were developed and maintained by private, municipal, or state agencies without benefit of federal standards for road construction. In urban areas such as New York, San Francisco, and Portland, Oregon, roadways were paved either with bricks or cobblestones, while in rural areas they were left as unpaved dirt roads. The search for a smooth road surface was of great concern to bicyclists until asphalt appeared for the first time in Paris in 1854. Tunnard and Reed state that "asphalt streets . . . were not seen in any quantity until the eighties and nineties, when bicycle enthusiasts forced improvement."[13]

The major legacy of the League of American Wheelmen was an effective lobbying organization that promoted the good roads movement at both national and state levels. In 1885, the League began to produce a free weekly publication entitled *The L.A.W. Bulletin.* The *Bulletin*'s principal editorial policy was to publicize national bicycle events and rallies sponsored by various local chapters around the country. In addition, it kept readers abreast of the international good roads movement and international bicycle events. Prior to the appearance of this publication, the L.A.W. had encouraged its members to publish articles about bicycling and the good roads movement in national periodicals of the day. Many state organizations had their own separate publications which they furnished to local members in addition to the *Bulletin.*

The first bicycles were so expensive that only a few could afford them, and early club members were for the most part prominent, white male citizens. This membership bias was a national

FIGURE 6
Rose Parade, Portland, Oregon

Rose parade, Portland, Oregon, turn of the century.
(Courtesy of Oregon Historical Society)

State Capitol, Salem, Oregon

State capitol, Salem, Oregon, turn of the century.
(Courtesy of Oregon Historical Society)

policy of the club, whose bylaws (appearing in the July, 1885 issue of the *Bulletin*) stated that "any amateur white Wheelman of good character, eighteen years or over with endorsement of two members . . ." was eligible for membership in the club.[14]

The Wheelmen were able to win support for the good roads movement in many states as well as at the national level. In 1899, the Good Roads Club was the first to create effective publicity for better roads in Oregon, by working hard and persistently lobbying for Oregon's Highway Improvement Bill of 1896. In addition, Oregon bicyclists were responsible for the adoption of the state's first highway tax. The tax later became the foundation for financing Oregon's highways.[15] Not all legislators, however, were as sympathetic to the Wheelmen as those in Oregon. In Massachusetts, one legislator felt that the League was a group of "idle rich" men who wanted to improve roads purely for selfish reasons; he withdrew support from a road improvement bill in that state when he learned that it was supported by the League of American Wheelmen.[16]

Regardless of the League's membership policies, anyone could propel a bicycle. Smith believes that the last social consequence of the bicycle movement was its liberating and mobilizing effect on the American woman. The new freedom which the American woman of the 1890s gained through the use of the bicycle was partially reflected in bicycle fashions of the day.[17]

> Generally, women defended the new style by saying that the long skirts, corsets, and heavy materials demanded by fashion not only endangered their health but had "recklessly cursed the unborn." Some women maintained that if they did not seize the opportunity to change to "rational dress," they had no right to ask for other privileges that had long been denied them. On the other hand most men were prepared to enjoy the sight of a neatly turned feminine ankle, but they became slightly apprehensive when the dress reformers maintained that the time was ripe for more radical changes.[18]

As more women bicyclists began to share the American roadways with their male counterparts, many of the Victorian male and

FIGURE 8
Independence Day Parade

Independence Day parade, Silverton, Oregon, turn of the century.
(Courtesy of Oregon Historical Society)

female mores that had governed social customs also disappeared from the American scene, along with skirts, corsets, and multi-layered petticoats.

THE DECLINE AND GROWTH OF THE BICYCLE MOVEMENT

In the summer of 1899, owners of electric automobiles were demanding that, like their bicycling counterparts, they be admitted to Central Park in New York City. This resulted in the organization of an American automobile club, and it is ironic that a similar event ten years earlier involving the bicycle had spurred establishment of the New York City Bicycle Club.

There were other indications that the so-called Golden Age of Bicycling was coming to an end. *Outing,* a publication that had begun its career as a publicity organ for the bicycle, stated in 1900 that the future would belong to the automobile.[19] The publication reported other indications of collapse in the bicycle movement, including a large controversy which arose over control of bicycle racing areas and which crippled the League of American Wheelmen, causing a rapid decline in membership. The League report of 1905 revealed that its national membership had dropped to 300. This severe decline led to the disbandment of the League by 1924.[20]

Further blame for the bicycle's decline was placed on the bicycle trust, the American Bicycle Company. Prior to organization of this company in 1900, there were ninety-five bicycle manufacturers, and their salesmen actively engaged in selling and publicizing the bicycle. However, with formation of the trust, the emphasis shifted from sales to management for there was no need to emphasize sales in a near-monopoly. Establishment of the trust led to the dismissal of many bicycle salesmen as well as a significant reduction in the advertising carried in various bicycle publications. The League of American Wheelmen was forced to stop publishing the *Bulletin* because of the drop in advertising, and this brought the demise of many local bicycle clubs that were dependent upon the publication for news of bicycling events around the country.[21]

It is unclear why the European bicycle clubs continued to thrive. Possibly the automobile was not as accessible to middle-class

Europeans as it was to their American counterparts. In Great Britain, the Cyclists' Touring Club (C.T.C.) noted that a decline did occur within the British bicycle movement after World War I.[22] During the depression of the 1930s, however, there was a bicycle revival.

> One newspaper reporter, L. H. Robbins, said that Americans had taken once again to the cycle because it was the most convenient means at hand to express their biological and spiritual dissatisfaction with the machine age. In 1936, over a million bicycles were sold. The bicycle appeared in New York's Fifth Avenue for the first time since the Gay Nineties. Ominously, the number of bicycle-related accidents climbed also, transcending five hundred a year in fatalities and fourteen thousand injured.[23]

The revival led to a reorganization of the League of American Wheelmen in 1934. Various European bicycle clubs, such as Algemene Nederlandes Wielrijders Bond (A.N.W.B., a Dutch club), Bicycle Federation of Belgium (B.F.B.), and the Cyclists' Touring Club, all noticed a substantial rise in membership during the 1930s.

Throughout World War II, the bicycle was seen as a pragmatic, immediate solution to the problems of gasoline rationing and rubber shortages in this country and abroad. Membership in the British Cyclists' Touring Club had fallen to a low of 25,000 by the end of 1940, and most people who knew about British bicycling expected it to continue to fall. However, the club actually began to revive, and by the end of 1944 it had grown to 36,000 members.[24] Another reason for the increased interest in Britain and other parts of Europe at this time was a change of emphasis in bicycle use from commuting to recreation. This trend has continued in the United States, and in this country since 1971 there have been more bicycles sold than automobiles—a first in this century. In fact, new bicycle sales within the United States have more than doubled since 1965, and tripled since 1960. Sales for 1972 were estimated by the Bicycle Institute of America to be 13.7 million, an increase of 65 percent over 1971 sales.[25]

With the number of U.S. bicycles doubling in the last decade, the question arises: Who is riding them? According to recent national studies conducted by R. M. Cleckner and F. F. Swim, 85 percent of bicycles in this country are used by children under the age of fourteen, but the proportion of young riders is declining rapidly.[26] The increase in bicycle use stems from a growing interest on the part of older people. This trend is reflected in the annually increasing number of adult models sold within the United States. In 1969, adult models comprised only 12 percent of the total sales, but by 1971, the Schwinn Company was aiming the bulk of its production at the adult market.[27] This spiraling trend in adult bicycle riding has been acknowledged by many metropolitan planning agencies, including those in San Jose, California; Portland, Oregon; and Atlanta, Georgia.

With their ever-increasing occupancy of American roadways, bicycles can no longer be regarded as merely recreational vehicles. The fact that bicycles have been outselling automobiles since 1971 should indicate that bicycles are a viable answer to commuter demands for transit modes which can provide personal transportation while alleviating the congestion, pollution, noise, and expense of automotive transit. In view of diminishing fossil fuels and natural resources necessary to build and maintain the automotive industry, it seems only logical that bicycles in conjunction with mass transit will become a much more acceptable transportation mode.

Automobiles are likely to continue for many years as the major form of long-range personal transit. It would be foolish to assume that a society arranged physically and socially around access to some means of automotive transport would change quickly. Even if the transformation were to happen quite suddenly, there would be an insufficient number of integrated, well-coordinated transit alternatives from which to choose. In view of this, the most pressing responsibility confronting planners and designers of transit systems is formulation of transitional alternatives to the present transportation network.

Historically, bicycles and bicyclists have been involved in the development of improved transit systems. Now it has become incumbent on the many new bicyclists, as well as all other persons using the transportation system, to become involved with the environment and insure its protection. Because of their excellent economy, flexibility, and environmental soundness, bicycles present planners and designers with an ideal mode, easily incorporated into existing land-use and traffic patterns. The direction and manner of organization chosen by people who are concerned with the future of transportation planning and the environment will in large part determine their effective impact. Organization is essential to the quality of responsive transit. It is only with a well-conceived planning concept that anything more than a purely cosmetic solution to bicycle transit and transportation problems will be achieved.

THE BICYCLE CELL

Unlike many other forms of transit, bicycles do not require expensive facilities for operation; however, systems like subways, monorails, trams, street cars, and electric buses force dramatic revision of land-use patterns. As self-propelled and self-contained units, bicycles operate independently of cables, tracks, or electrical wires. Thus they are a transit mode which can be integrated with minimal environmental impact. In fact, they can be joined with other modern systems; for example, if special provisions are made, riders can bring their bicycles onto the subway, car, or bus.

In the "dual-mode" or bikes-along method the rider brings his bicycle with him on the public transit vehicle. In a properly designed system of this type the bicyclist could ride right up to the vehicle, load his bike quickly and easily on board, ride to the exit station, and then disembark and bicycle to his destination. Assuming near-level terrain and minimum delays, the cyclist's average speed should be at least 10 mph . . . the dual-mode trip can be fast enough to be competitive with automobile

travel. Use of a bicycle permits the rider to pedal to
within several yards of his destination, avoiding all the
difficulties of parking a car.[28]

The competitive aspect of this concept is merely a restatement
of the idea that alternative transit modes are available. As Sommer
and Lott state, "To neglect any system is to overload others beyond
their optimal carrying capacity and thereby to do an injustice to
all."[29]

Today, most bicycle travel in the United States is focused
within local neighborhoods, or *cells,* according to surveys conducted by municipal and private agencies. A cell is a distinct geo-
graphic area within a community which has common transportation
needs. A cell is defined through physical separation, census tract
data, or distinguishing features such as business, education, recreation, or industrial use. The people working and living within a cell
can be thought of as a distinct group.

One of the most prominent groups of bicycle users within the
cell consists of children between the ages of five and fourteen years.
In addition to recreation, they use bicycles for commuting, mostly
between home and school. The second largest group of bicyclists—
and the most rapidly growing one—is comprised of adults who ride
for recreation and who commute between home, school, and civic
and commercial centers. This second group has increasingly used
the bicycle for travel between cities and from one region to another.

If bicycle planning is to be responsive to bicyclists' needs and
desires, it must begin at the cell level. Past experience has shown
that bicycle planning which places primary emphasis on city-wide
rather than cell networks faces problems in both acquisition of
funding and adoption of the bicycle network by city agencies.

Cell boundaries can best be established by local planning agencies, on the basis of elementary school districts, voter precincts, or
boundaries of neighborhood planning associations. Where neighborhood planning associations do exist, bicycle planning need not be
coordinated as a secondary function of the association. Rather, the
organization can be extremely helpful in designating the bicycle
planning cell and, once it has been established, in helping to form a
citizens' task force.

This task force would investigate the needs and desires of people involved in the neighborhood—homeowners, renters, workers, and business people—and then develop a preliminary plan for bicycle travel in the cell. The task force would coordinate its efforts with city or regional planning agency representatives in order to integrate its plan with those of other cells.

Conceivably, some neighborhoods may have little or no interest in bicycle planning and may choose to leave it up to the city planning agency. In this case, the city planner would develop a preliminary plan for bicycle transit within the cell and then present it to residents for comments and approval. Since bicycle cells in most cases will have no legal authority or regular funding, and thus will operate as voluntary organizations, it is probable that interest in planning will vary from cell to cell. The cell concept proves its value, however, in the fact that none of the cell members need feel obligated to become involved, yet organizational channels permit any number of persons within the cell to participate. In the event that only one or two people are interested, they could conceivably develop the complete neighborhood bicycle plan themselves.

Until now, bicycle planners have relied upon questionnaires of varying ambiguity to solicit public input. The complexity of these questionnaires, coupled with the public's unfamiliarity with the problems of bicycle planning, has resulted in planning data that is speculative at best. Furthermore, the majority of surveys have generated little public interest in bicycle planning, and often responses have been incomplete.

An alternative to the traditional questionnaire is the Cycle Pack, a graphic information package which we devised to capture public interest and elicit more complete information for planners. The Cycle Pack is directed not only at bicyclists, who represent a relatively small segment of the community, but also motorists, homeowners, and all others who live and work in the cell. If survey questions are directed at all of these interest groups, the respondents will feel more directly affected by the outcome of the survey and will be more inclined to respond fully. It is important that the Cycle Pack be distributed at the cell level, for the localized nature of bicycling calls for neighborhood input into the planning process. By beginning with a high-resolution plan based at the cell level, the planning effort can retain a user-oriented approach when it is expanded to the city, inter-city, and regional levels.

The Cycle Pack's design depends on the nature of the particular cell surveyed and the type of information desired. It might include a map marked with the cell's prominent geographical features, historical points of interest, and local recreation areas, as well as marking pens, instructions, and simple questions. If the cell is based around an elementary school, for example, the Cycle Pack might direct students and their families to color in the bicycle routes which they ride or would like to ride. Their proposal of future bikeways could be especially helpful to later design and engineering efforts if the respondents give brief explanations as to why they chose these new routes. The kit could also provide suggested solutions to local bicycle transit problems and new ideas for travel within the cell. In addition, there might be information on bicycle safety.

The task force could distribute a similar version of the Cycle Pack to cell residents not reached by the school, such as senior citizens. Older people within the cell are often most involved with the neighborhood, can provide valuable insights to its history, and can offer a helpful, general perspective. This type of information is extremely important, as the test of a bikeway is whether it truly reflects the intrinsic characteristics of the neighborhood.

Cells based in industrial or business districts would receive basically the same type of Cycle Pack, but with slight modifications in their instructions. Problems of specific concern to a business or industrial district—for example, large rapid transit terminals and main corridor intersections—could be explained in greater detail by the task force or by the assistant from the city planning agency.

A PLANNING MODEL

The following is a general model which illustrates how a bicycle plan can be implemented at the cell level. It should help to clarify the process by which cell organization could be extended to city, inter-city, and regional levels.

In a community where attitudes and values are favorable to supplementary modes of transit, neighborhood bicycle cells could

Planning Structure

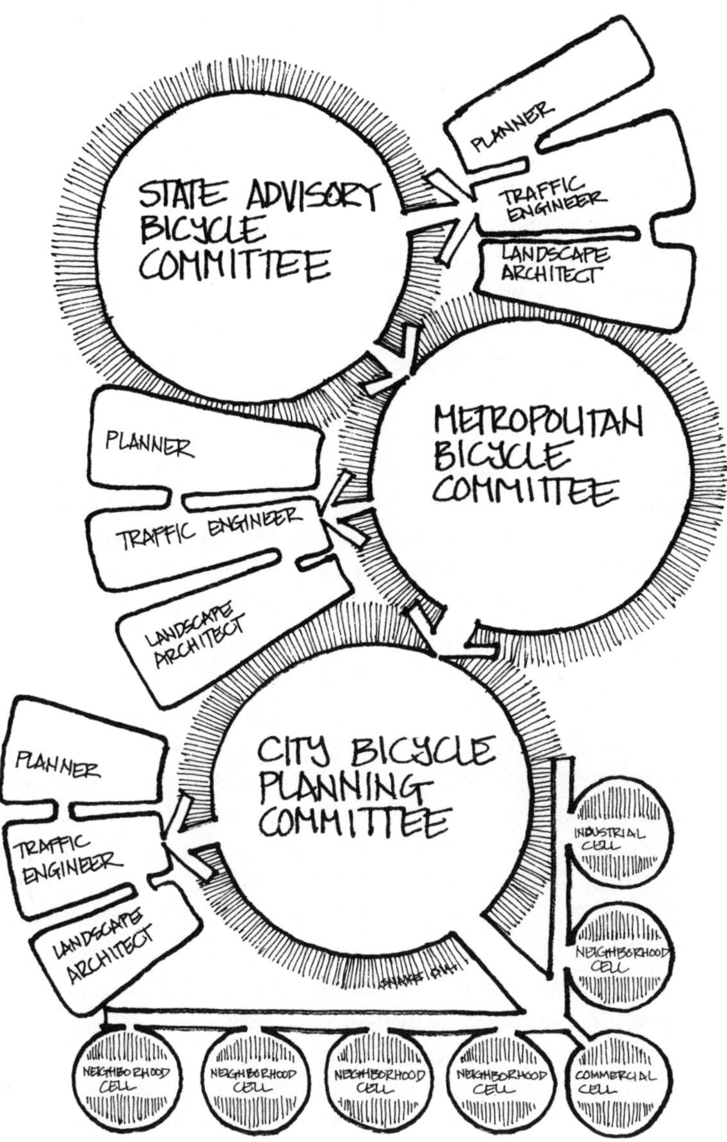

Planning structure based on bicycle cell.

be established. The city planning staff or agencies in charge of setting boundaries for school districts, voter precincts, or neighborhood planning associations should define the parameters of the cells.

Once cells have been designated, the city planning commission assigns a staff member to help organize a citizens' task force for each cell. The task forces change their memberships at regular intervals and their governing boards should consist of no more than seven members.

If cell residents show no interest in formulating a bicycle plan, the city takes responsibility for developing a preliminary plan. The proposed plan is then subject to approval of bicycle cell members.

The first action of the citizens' task force is to determine the extent and type of bicycle use within the cell. After initial investigation, the task force distributes Cycle Packs to all persons working and living in the cell. These packs may be handed out through the elementary school or by members of the task force.

After a specified amount of time, the Cycle Packs are returned and the task force reviews them and makes recommendations to the city planning agency. All recommendations are ratified by a specified consensus of the task force. If the city agency rejects the preliminary bicycle plan, it is returned to the task force for revision.

Once a bicycle cell plan has been adopted by the city planning commission, it is forwarded to the city council for final approval. If the city council approves the proposed bikeway network, it then allocates funds for implementation. Recommendations for city council funding should be discussed in a general meeting of the task force and all concerned citizens. Basically, all cells should be subject to equal allocation of funds; however, cells with significantly higher populations may receive proportionately more funds and staff time.

Upon ratification of individual cell plans, city planners begin to integrate the cell plans into a comprehensive city plan. Task forces and planning agencies make the final decisions regarding nodes of intersection with other bicycle cells.

Representatives from each neighborhood task force serve as members of a city-wide bicycle planning committee. Membership in this city-wide board should probably not exceed seven, and no individual should serve on more than two bicycle committees at the neighborhood, city, metropolitan, and regional level. The city-wide committee is advised by members of the city planning agency; staff advisors should include traffic engineers, landscape architects, and planners.

Once the city bicycle planning committee has formulated a comprehensive preliminary plan to integrate all existing urban corridors and cell plans, it submits the plan to the city planning commission for adoption. After it is approved, the city-wide preliminary plan is handed to the city council for ratification. Any plans which involve bicycle routes connecting neighborhood cells with central commercial and business districts should be worked out by the city planning department in cooperation with the central business district cell, neighborhood cell, and city bicycle planning committee.

Having completed a comprehensive bicycle plan for the city, planners then consider an inter-city or metropolitan plan. Members of the city bicycle planning committee are appointed to positions on a metropolitan bicycle planning committee. The metropolitan committee then works with members of all city planning commissions with the metropolitan transit district, formulating adjoinment and extensions of inner-city systems into a metropolitan plan.

Inter-city planning which connects bicycle systems in a metropolitan network is handled through regional planning agencies. In areas where they exist, county agencies with the capacity and jurisdiction to coordinate the metropolitan plan work with city planners and members of the metropolitan bicycle planning committees.

Creation of an inter-metropolitan bicycle transit network can be handled through a state bicycle advisory committee comprised of members of metropolitan planning committees. The state advisory members work with state transportation and planning committees. Allocation of state funding is regulated by the state advisory committee and state funding handled through legislative recommendation. Federal funding is similarly controlled and distributed by the state advisory committee, the state transportation department, and the state legislature. The state bicycle advisory committee may also help to formulate uniform bicycle traffic control devices.

Every attempt must be made at the initial stages of design to foresee problem areas and to plan adequately for them. Future construction of schools and freeways or any other major land-use changes will require modifications of the area's bicycle facilities.

Failure to plan for these changes in advance will result in crisis planning with scant time for research and design. A general policy should be developed which can act as a basic guideline for planning in emergencies.

CONSIDERATIONS IN CREATING BIKEWAYS

Although it is nearly impossible, as well as undesirable, to develop a textbook approach in establishing aesthetic and experiential criteria for bikeway designs, it is possible to formulate goals which may act to increase the designer's awareness and sensitivity. Hopefully this will counteract the overwhelming tendency of designers to depend on engineering and statistical data in the conception of bikeways.

Many natural features and spaces within a neighborhood cell afford designers a variety of experiential situations. By closely examining the essential quality of physical features, views, smells, sounds, and so on, we can gain some notion of what is meant by "experiential" and can understand how it pertains to the design process. Perhaps the designer's greatest job is to establish an intimate rapport with the place in which he or she wishes to formulate landscape additions or modifications. It is only when the designer gains a sensitivity to a particular place, especially over different periods of time, that he or she can intuitively create places in which people will respond with continual diversity. People's ability to find new experiences in the same place many times is a testament to the success of design.

Take an alleyway within an older suburban neighborhood, for example. Alleyways are often rich, diversified places. They expose the intimate, very special milieux of people's back yards, while at the same time may contain an overgrown trailer loaded with tires and peeling cans. The visual diversity created by the combination of a well-kept patio and flower garden and a bramble-covered trailer is extremely important, in that it establishes an element essential to design: contrast. By designing bikeways through areas of existing contrast, or by designing contrast into a bikeway, the designer insures people the opportunity for variation and choice. Presenting this opportunity in essence allows people to create their own personal relationships with a place. Changes in scenery involving waterways, spaces between buildings, shade and light, and color in all of its forms and arrangement through the seasons, and the various textures of the space, all coalesce as potential stimuli for choice.

Certain areas along a highway or within a city always seem very explicit in the responses which they conjure up from passersby. A freeway rest stop may never appeal to some people because they feel intimidated by the possibility of encounters with other people or situations with which they feel incapable of dealing. People's needs and feelings for interaction, seclusion, and the whole range of experiences between the two fluctuate constantly and are very often situational; thus a variety of locations which satisfy these changing needs and feelings are necessary. Public places which invite people to freely interpret and use them, as well as places with specific functions like rest and picnic areas which demand a more defined response, all have value within an experientially diverse bikeway.

Bicyclists most often travel in groups of two, although often one person travels alone. In bikeway design, then, careful attention must be paid to scale and context of areas created for people's use. A space created for interaction should be accessible to other places that allow people to remove themselves from the group altogether or to participate by watching, remaining secure in the fact of their separation. While it might seem more practical to build a rest and picnic area to accommodate fifteen people with benches and tables, perhaps designing places for groups of two or three and several spaces suitable for only one person might be a better solution. It is difficult to determine what types of places belong where, without the given context of neighborhood, city, region, or geography. A discerning eye for the proper contextual framework will enhance the function which a place is designed to perform. It must be remembered that the route selection process begun with the Cycle Pack is also the beginning of the design process. Planning, design, and engineering are mutually dependent components of the same process.

People often avoid new experiences because they fear they will disrupt their orientation and place themselves in a position of physical and mental jeopardy. This intuitive avoidance of new situations can be alleviated in part by designing bikeways and areas along bikeways so that anxiety-causing elements are removed. Narrow bikeway bridges, elevated urban bikeways, or bikeways which pass

FIGURE 10

Elevated Urban Bikeway

Elevated urban bikeway. (*Milwaukee Stadium Freeway, Wisconsin*)

through tunnels should be designed so that the greatest amount of information concerning the special situation is made obvious to the user. Furthermore, the use of materials and plants indigenous to the area will give people a sense of natural territory and will help to reinforce the bikeway user's feeling of security.

Priorities in the design of individual neighborhood bicycle cells will be somewhat different depending upon the population makeup of cells and the criteria used in designating them. Residential cells will most likely place highest priority on routes to schools, recreation areas, and mass transit terminals within the cell. Commercial cells may want to focus on distribution networks adjoining major bikeways, while industrial cells may center their efforts on one or two routes which are heavily used or serve peripheral automobile parking lots. Bikeway designers, whether dealing with residential, business, or industrial areas, must remain aware of the necessity to combine a diversity of elements within any bikeway design. All designs should contribute to the continual interaction of people, in a variety of situations and spaces.

BICYCLE CORRIDORS

Bicyclists should be considered neither as motorists nor pedestrians, but rather as a unique transit group with special requirements. This must be kept in mind when future bikeway corridors are being selected and evaluated. The route selection process initiated with the Cycle Pack is the most direct means of gaining satisfaction of these specific needs.

Bikeway corridors may be planned to accommodate pedestrian, bus, automobile, and light rail traffic. Most often, bikeways will be located in conjunction with one or more of these transit modes, and therefore selection of corridors to accommodate these other forms of transit represents pragmatic and efficient land use. However, if it can be demonstrated that a particular piece of land is needed only for solutions to problems of bicycle circulation, this should be accepted as valid and necessary land use. Extra land costs will be compensated for in reduced congestion of automobile arterials, increased recreational and aesthetic enjoyment, and a greater degree of safety for all transit modes within the immediate network.

Public and private planning agencies throughout the country have evolved a number of different classification methods for bicycle arterials, with semantic confusion the result. Terms such as *bikeway, bike-lane,* and *bike-path* are commonly given different meanings by different planning agencies. For the purposes of this book, bikeways are described according to the following three major classifications.

Class I bikeways are designed for the exclusive use of bicycles and are completely separated from automobile and pedestrian traffic. Crossings with automobiles and pedestrians are kept to a minimum or are avoided through the use of over- or underpasses. By nature, Class I bikeways will occur most often in open spaces, parks, highway rights-of-way, railroad rights-of-way, river and canal banks, and newly planned developments.

FIGURE 11

Class I Bikeway

Class I bikeway: for the exclusive use of bicycles.

FIGURE 12

Class II Bikeway

Class II bikeway: adjacent to, but separated from, automobile and pedestrian traffic.

Class II bikeways are adjacent to but separated from automobile and pedestrian traffic. Bicyclists are provided a separate path, with the exceptions that it may be crossed by parking automobiles or automobiles turning into or out of driveways, and it regularly crosses intersecting streets. Class II bikeways are segregated from automobile and pedestrian traffic with a physical barrier or by painted markings. They are one-way, occurring on both sides of two-way streets or on the right side of one-way streets. This type of bikeway will contribute the most to developing the bicycle as a supplement to the automobile for commuting and transportation in existing developments. This is because they can be incorporated as part of the overall city traffic pattern—they share common traffic corridors with other transit modes.

Class III bikeways are shared bikeways in which the bicycle occupies the same right-of-way with either automobile or pedestrian traffic. Signs designate the road or sidewalk as a bikeway, warning automobile drivers or pedestrians that bicycles should be expected.

FIGURE 13

Class III Bikeway

Class III bikeway: a shared bikeway in which the bicycle occupies the same right-of-way with automobile or pedestrian traffic.

Although this is by far the least expensive type of bikeway, it is also the least safe for the automobile, the bicyclist, and the pedestrian. The Class III bikeway should never be used on high-speed, congested roadways or on heavily used sidewalks, and is recommended only for temporary purposes. For a heavily used road, a Class I bikeway of some sort is required because complete separation is the only means of providing real physical safety as well as a feeling of protection in the bicyclist. One solution which provides this safety as well as the ability to cross freeways and their access ramps was developed in the study *Milwaukee Stadium Freeway*. The concept involves completely elevated bikeways.

THE ELEVATED URBAN BIKEWAYS are conceived as being an isolated, elevated freewheeling structure, dipping and weaving and leaping along in the freeway right-of-way, ducking under flyovers and bridges at grade. In this manner the urban bikeways would link together various communities on opposite sides of the freeway. The cyclist would get exciting views of the adjacent scenery and points of interest, as well as dynamic peep views of the surging streams of motor traffic below. . . .

The isolated URBAN BIKEWAY STRUCTURE would be developed in a UNIT SYSTEM set on piles.

FIGURE 14

Elevated Urban Bikeway Plan

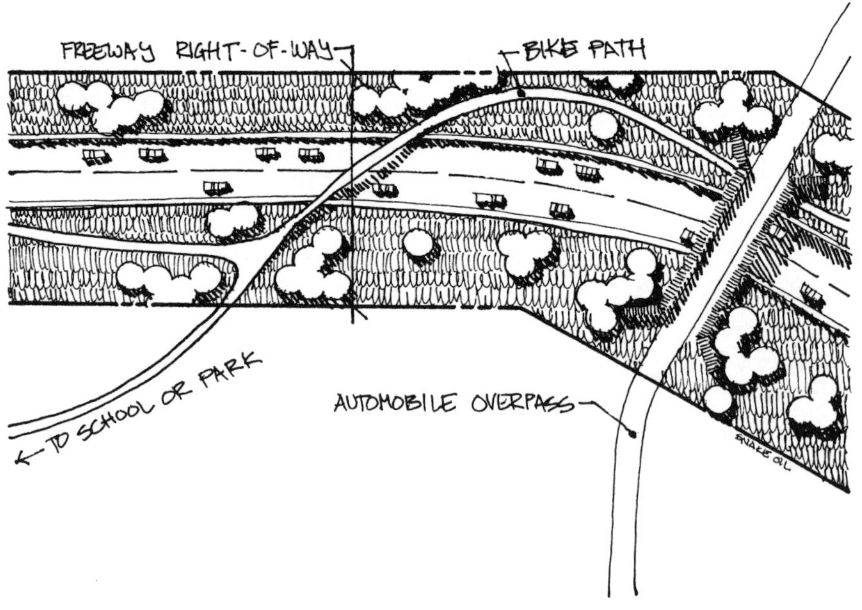

Plan for elevated urban bikeway. (*Milwaukee Stadium Freeway,* Wisconsin)

Units would be made in precast concrete or lightweight
pressed metal sections. These units would sit on top of
the pile column supports. The unit sections could be
MASS PRODUCED and marketed throughout the many
cities with freeway systems in North America. Thousands
of miles of URBAN BIKEWAYS could be implemented
on a national scale, thus opening up the cities of the nation and giving benefit to millions of people.[30]

The conventional Class I bikeway may be more suitable in other
situations, but crossings of access ramps and the freeway itself
should always be separated, if only by integration with existing
automobile over- or underpasses.

The following are all examples of Class I bikeways.

Abandoned Railroad Rights-of-Way

The decline of both passenger and industrial railroads in the
United States has left a large number of railroad-owned rights-of-
way lying unused. Abandoned rights-of-way connect towns, cities,
and industrial centers within large urban areas, but for the most part
are presently being used for recreational purposes. Rights-of-way
offer excellent commuter possibilities, yet planning agencies and
railroad officials are reluctant to develop or sell these lands, as they
have the potential to serve as urban mass transit corridors in the
future.

One important value of railroad rights-of-way as bicycle transportation corridors is that usually they consist of property owned
by only one company or governmental agency and can therefore be
utilized without the complicated proceedings needed to acquire
many small parcels of land. Also, this property is linear in form and
virtually unbroken, usually stretching between cities and towns or
between activity nodes within the city. Furthermore, original selection of the corridors was highly dependent upon their maintenance
of fairly constant and gentle grades, which is also a major requirement of bikeways. Abandoned railroad rights-of-way are often quite
scenic, running through wooded or undeveloped areas. Where they

FIGURE 15

Abandoned Railroad Right-of-Way

Abandoned railroad right-of-way, Tiburon Peninsula, Marin County, California.

run through urban areas, they could be developed as linear parks which lend softness and identity to the urban landscape.

This type of bikeway would be an excellent example of the bicycle serving as a transitional mode between the automobile and future, more efficient rail transit. Transformation of the railroad right-of-way into an inexpensive bikeway which might cost as little as paving a one-lane road could make possible both recreation and commuter use by bicyclists until a rail system for the route became a reality. This approach would keep the ever-increasing number of bicyclists off the heavily used automobile arterials until a completed rail system could draw large numbers of automobile commuters from the streets at rush hours. The street systems could then handle a larger number of bicyclists and the railroad right-of-way bikeway might not be needed. This plan assumes that the railroad right-of-way cannot handle both rail transit and a bikeway.

A State of Indiana study set up criteria for the evaluation of particular railroad rights-of-way as bicycling and hiking trails. Though they were devised for recreational purposes, these criteria apply nearly as well to commuter bikeways.

PHYSICAL MEASUREMENTS: Right-of-way should be a minimum . . . width of fifty feet. Developed riding surface should be eight feet wide to accommodate motorized vehicles for maintenance.

AESTHETIC QUALITIES AND PHYSICAL FEATURES: Pleasant and changing. . . .

TOPOGRAPHY: Gently rolling, presents various challenging areas to the rider or hiker with the total grade not exceeding 10 degrees. Adjacent areas to the track could have extreme elevation changes to provide a more enjoyable experience for the user.

POINTS OF INTEREST: Historic areas, scenic overlooks, and other significant points should exist along the trail for providing interest and identification of riders' location along the trail.

PROXIMITY TO SERVICE FACILITIES: Since most railroad lines run through small communities, . . . service facilities should, in most cases, be available. Service facilities should be available to the riding or hiking public every six to ten miles along the trail.

ACCESS POINTS: Although access to the trail will usually be available at every road crossing, the users of the trail should be encouraged to enter the trail at specific access points. By designating specific points of access, there may be an elimination of possible encroachment on private property as well as preventing possible congestion at busy road crossings.

POTENTIAL CONNECTION TO OTHER TRAILS: In proposing a trail development, consideration should be given to connecting with other trails that might be located in the area. This consideration should eliminate any duplication of trails and promote the establishment of a more usable and acceptable trail system.

PRELIMINARY WORK REQUIRED FOR DE-
VELOPMENT: All potential rights-of-way should be in-
vestigated for current conditions and work that may be
required before development of public trails can be
initiated. The existence of railroad ties, rails, bridges, extensive undergrowth, erosion, and other problems should
be documented for each trail segment.[31]

In addition to providing pleasant commuter routes to work
and to shopping areas within the city, the railroad right-of-way can
serve as a bicycle and pedestrian link to regional recreation areas on
the periphery of the city. The proposed Cross Marin Trail and
Bicycle Route is one such system; an old, abandoned, narrow-gauge
railroad right-of-way is used for part of the route.

The route begins and ends at two significant landmarks in
the County. At the east end is the Marin County Civic
Center, which is the hub of civic and cultural activity and
is a major urban center in the County. This landmark is a
natural regional beginning to the route in that it is both
distinguishable and accessible and it also has available
parking for persons using their automobiles to get to the
route. It is also possible that bike, buses, and/or internal
transportation in the future could accommodate a larger
number of users from this point.

The west end of the route is a natural termination in
that it joins the edge of both the Point Reyes National
Seashore and the newly created Golden Gate National
Recreation Area. These areas will be the major desti-
nations for bicycle riding activity in that the user can
arrive and explore in an atmosphere relatively free of
automobile intrusion, pollution, and noise and can actively participate in a healthy recreational sport amid the
beauty of the region's largest parks.[32]

Canals and Waterways

The use of canals and waterways as focal points for linear recreation spaces and for bicycle and pedestrian transportation routes
is increasing.

FIGURE 16

Bikeway Canal Route

Bikeway canal route, Kentfield, California.

Canal banks are attractive resources for the development
of bike paths. The grade is usually just enough to keep
the water flowing and from becoming stagnant. Aesthetically it can be a pleasurable place to bike, particularly if
there are small oasis-like rest areas located along its
course.[33]

The City of Wichita, Kansas, has proposed the Canal Route
Open Space Corridor as part of Wichita's *Open Space, Parks, and
Recreation Plan.* The corridor would contain facilities for riding,

bicycling, and hiking trails, roadside park facilities, and water-based recreational facilities. The plan recommends that "the local governing bodies recognize and further use the existing river and creek systems and the Wichita-Valley Center Flood Control Project as the logical framework of connecting links for the entire open space, park, and recreation system."[34]

The concept of using the Canal Route as an open space corridor rather than simply building a standard freeway system along the route arose from a desire for more sensitive planning and design of Wichita transportation systems.

FIGURE 17

Bikeway Canal Route/Open Space Corridor

Bikeway canal route/open space corridor proposed for Wichita, Kansas.

Over the past few years, there has been a growing concern
locally and nationally over the impact of urban highways
on neighborhoods, social patterns and environmental
quality. In response to the questions raised, many cities
have instituted design teams to integrate all aspects of
highway planning. The engineer is now likely to be joined
by architects, sociologists and economists, as well as park,
school, and city planners in an effort to make the traffic-
way a more compatible part of the urban environment.[35]

With the establishment of a network of greenways, or open
spaces, along canals, which contain Class I bikeways and pedestrian
paths, a city-wide bikeway network can exist with very little con-
flict with automobiles. Short links on low-use streets can then connect this system to the more conventional Class II networks of
residential and commercial areas. The greenway network can then
function both for recreational and commuter traffic.

New York State's Erie Canal Bike and Hikeway between Utica
and Marcy is a 5.125-mile recreation bikeway, the first section of a
much longer bike and hikeway along the Erie Canal. This is one of
a number of recreation areas whose focal point is a linear body of
water. In this case, the waterway is an active commercial barge
canal; in other cases the channel may no longer serve industrial purposes. Many of these water routes are merely drainage canals or old
river channels.

The Paseo del Rio in San Antonio, Texas, is a development
which follows an old river channel through the city's highly developed core.

. . . either way you look there is a pleasant view. Trees
shade the banks, cypresses that turn rusty with fall. An
oak arches above the water. The river curves, each way,
between walks bordered with kept grass, adorned with
cannas, banana plants, and bright flowers in season. . . .
in the noise of traffic, among the shifting figures crossing [Commerce Street Bridge], you may not be aware
. . . all this lively scene, only the river itself. . . .[36]

The Paseo del Rio has no formal bicycle facilities but is an example of what can be done with a river channel in an urban area.

FIGURE 18

San Antonio River Channel

San Antonio River channel (Paseo del Rio), San Antonio, Texas.

It has become a focal point of the city: "shops and restaurants which once turned bleak, inhospitable backs to the river have faced about, adding the cheerful animation of the bazaar and the sidewalk café to the leisurely pace of the promenade."[37]

Highway and Freeway Rights-of-Way

The use of highway and freeway rights-of-way for inter-city and state-wide bikeway systems is a promising possibility because these lands are already owned by governmental agencies, and connections to other bikeways and low-use roads are usually simple and direct. One major obstacle in the use of these corridors by bicyclists, pedestrians, and mass transit is the attitude of government bodies in charge of transportation in these corridors. Often city, regional, and state agencies all can be blamed, though the latter has been by far the most reluctant to consider systems other than the automobile as legitimate means of transportation.

In its recent study, *Asphalt Empire,* the Oregon Student Public Interest Research Group (OSPIRG), criticized the Oregon State Department of Transportation for ignoring the land-use implications of highways, not seeking adequate citizen input into long-range planning or individual projects, not seriously considering alternatives to highways, and not once using federal funds which have been available since the early 1960s for the construction of mass transit facilities.[38] The same study was also critical of the highway lobby, which consists of truckers, teamsters, auto clubs, oil and gasoline dealers, Associated Oregon Industries, and Associated General Contractors.

> The power highway lobby has also contributed to the state's abundance of highways and lack of mass transit. As one of the largest campaign sugar daddies in the state, the lobby has done its best to insure that highway construction dollars would continue to flow into its pockets. It has already set transportation reform back several years.[39]

The OSPIRG study was concerned mainly with the Oregon State Department of Transportation's attitudes toward development of mass transit systems, and this same concern has led critics in other states to ask similar questions about their own state transportation departments. In California, whose highway lobby is perhaps one of the most powerful in the country, automobile transportation planning has been the principal concern of the California Department of Transportation. With state transportation departments resisting so strongly the pressures for mass transit, can bicyclists hope for just consideration?

If a balance can be achieved by transportation planning agencies with regard to various transit alternatives, then perhaps corridors can be evaluated fairly and intelligently in terms of their most suitable use. If it is found that it would be advantageous to route a bikeway in a freeway corridor, some special consideration to safety and the enjoyment of the bicyclist could be given.

MIXED-MODE SYSTEMS

Mixed-mode systems of transportation are those systems which involve the bicycle and at least one other vehicle as a means

FIGURE 19

Automobile-Mounted Bicycle Rack

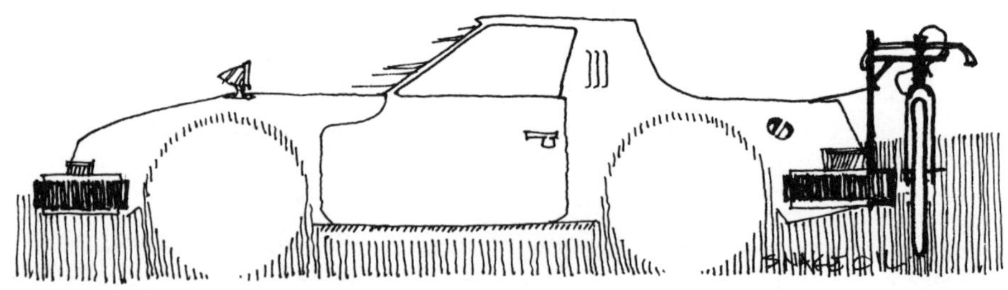

Bicycle rack mounted on automobile.

of transportation or recreation. The most common mixed-mode system is one in which bicycles are transported between recreation areas and home by an automobile.

Since the Bay Bridge between San Francisco and Oakland, California, has no bicycle facilities, AC Transit of Oakland is experimenting with a bicycle bus called the "pedal hopper," which allows bicyclists to bring their bicycles on the bus for the journey. Unfortunately, the experiment has been found financially unfeasible, even at fares of $1.25, and user participation has barely kept it going.

FIGURE 20

AC Transit's "Pedal Hopper" Loading

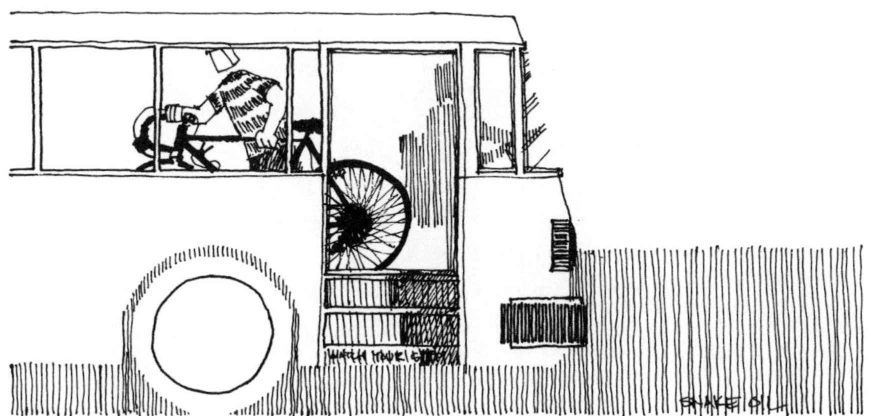

AC Transit's "pedal hopper," Oakland, California.

In the same region, the Bay Area Rapid Transit System (BART) has proposed a system of trails and parking facilities to provide bicycle access to its terminals. Further provisions for mixed-mode systems have been made in the Bay Area by the East Bay Bicycle Coalition. The Coalition has developed a rack which can carry up to nine bicycles on the back of a bus.[40] These systems, and others such as the "bike/ferry" system operated by the Golden Gate Bridge District, offer alternatives or supplements to long-distance bicycle commuting and conventional automobile commuting, as well as to existing and proposed rapid transit systems.

FIGURE 21

Bus-Mounted Bicycle Rack

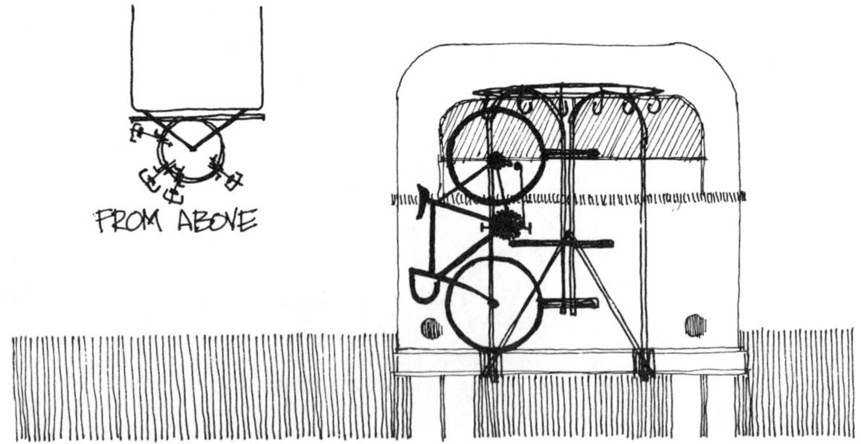

Bicycle rack for the rear of buses developed by the East Bay Bicycle Coalition, Berkeley, California.

Transporting bicycles to recreation areas by automobile has been developed to the extent that special equipment and facilities are sufficiently available, and this system has proved quite practical. For the minimum expense of an automobile bike rack, anyone with an automobile and a bicycle may participate. The following, more complete mixed-mode systems for commuter transportation, however, are for the most part still experimental.

Bicycle on Automobile to Peripheral Parking

The use of standard automobile-mounted bicycle racks and provisions for automobile parking outside concentrated employment areas is perhaps one of the simplest mixed-mode systems.

With respect to concentrated employment areas, mixed-mode provision could result in a shift in the transportation complex of downtown areas and central business districts. In such an eventuality it is easy to envisage increased automobile parking in the peripheral areas, with increased bicycle usage for transport into and within the centers. If the shift is sizable, this may significantly relieve parking and traffic congestion in the centers.

. . . A bikeway system network that criss-crosses an entire urban area may be less effective than a set of short-length bikeways within, and radiating out from, employment or other trip generation centers to peripheral automobile parking facilities.[41]

The importance of this mixed-mode concept lies in its ability to link bicycle districts with employment centers, particularly when extreme grades, bridges, and freeways that lack adequate bicycle facilities restrict direct bicycle connections.

Due to the simplicity of combined bicycle and automobile transit, this type of mixed-mode system may prove to be the most flexible and most responsive to changing urban transportation needs. Until some form of public transportation can accommodate both bicyclists and their bicycles as easily as the automobile, the automobile will undoubtedly remain the most common "other vehicle" in mixed-mode systems.

FIGURE 22

Peripheral Automobile Parking Facilities

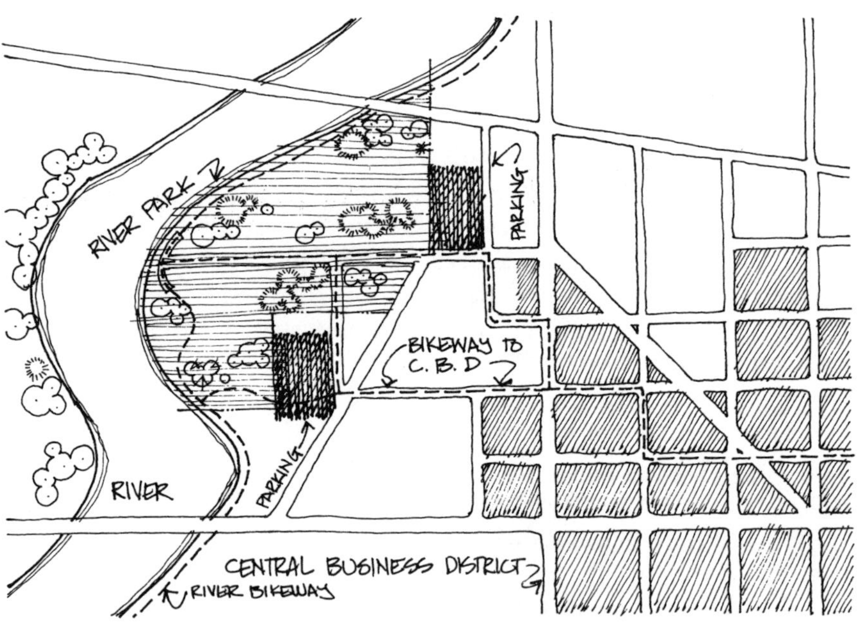

Plan for peripheral automobile parking facilities which could serve a central business district during the week and a regional park on the weekends.

Bicycle to and from Public Transit Terminals

When dealing with combined public transportation and bicycle systems, it becomes necessary to distinguish between two types of public transportation: bus systems and light rail systems. The distinction lies in the frequency and location of stops or terminals along the system. Bus systems have the potential for passenger pickup and drop-off at locations close to pedestrians' origin and destination points. Light rail systems are less able to provide for this degree of pedestrian accessibility in an urban area because they have fewer terminals with greater distances between stations. San Francisco's BART system is already experiencing major overcrowding at the automobile parking lots which adjoin the terminals.

As more cities turn to rapid transit systems like BART, the possibilities for the bicycle serving as an important transit mode to and from terminals becomes greater. Evidence shows that significant numbers of commuters are turning to the bicycle to solve this problem. Unless safe and convenient parking facilities for bicycles are provided at the terminals, however, the parking problems will merely be transferred from the employment area to the residential area.

A study conducted for the Atlanta Metropolitan Region recommends six design criteria which should be considered for accommodating the bicyclist at transit stations:

> Facilitate ease of access by providing signed bicycle approach routes. . . .
>
> Bicycle storage facilities should be located as close to station entrances as possible. . . .
>
> Where possible, bicycle storage facilities should be within view of personnel working at the station, or close to high activity areas to minimize the chance of theft. . . .
>
> At subway stations where no automobile parking is provided, consideration might be given to purchasing or leasing bicycle storage facilities. . . .
>
> Charging fees for bicycle storage is not recommended. . . .

> Consider converting automobile parking spaces to
> bicycle storage areas if bicycle patronage becomes
> high. . . .[42]

Bicycle on Public Transit

The most controversial mixed-mode possibility involves the modification of either public buses or rail vehicles to enable passengers to carry their bicycles with them. Long-distance systems already allow bicycles to be stowed with other luggage, but this is inconvenient and time-consuming and becomes impractical for short-distance or commuter trips. While European trains are currently being used successfully for both long- and short-distance commuter trips with bicycles in baggage compartments, the state of American railroad systems makes this all but an impossibility here. At present, the most promising systems in this country allow passengers to bring their bicycles into the modified passenger compartments of buses and rail vehicle cars. As previously mentioned, AC Transit of Oakland, California, has experimented with a "pedal hopper" service over the Bay Bridge. Basically, the interior of the coach was modified by removing seats between the front and rear wheel wells and bicycle racks were installed. Accommodations for twenty-four riders and their bicycles were provided in each coach.

As with the "pedal hopper," bicyclists commuting on light rail systems have run into difficulty; for instance, BART restrictions have limited bicyclists' use of the system. Until recently, BART allowed only folding or collapsible bicycles on the system[43]; as of January, 1975, a six-month experimental program permitted limited transport of regular bicycles on BART cars during off-peak hours (week days 9:30 am-3 pm and after 6:30 pm). Only bicyclists who obtained user permits from the BART administrative offices could participate in the program, and they could board only the rear area of the last car on the train.[44]

This type of mixed-mode system enables people to bicycle from home to residential area terminal and from the downtown terminal to work. Bicyclists have their bicycles with them at all times and are not forced to pay for parking at the terminals or for other

FIGURE 23

Interior of "Pedal Hopper"

Inside the "pedal hopper," Oakland, California.

transportation to and from terminals. People who need their bi-cycles for both the residence-to-terminal trip and office-to-terminal trip are not placed in the curious and economically unfeasible position of owning two bicycles and having to leave one at a terminal overnight and the other at a terminal during the day.

All of the above mixed-mode systems and their possible combinations will be of increasing interest to transportation and comprehensive planners. These and other, as yet undeveloped, mixed-mode transportation systems may prove to be valuable elements of a well-balanced transportation plan.

PLANTING

Plants are the most versatile tools available to planners and designers of transportation corridors. They separate and delineate various transit modes; effectively minimize extremes of hot, cold, wind, and rain; and screen and protect riders from the harsh glare of sun on concrete and glass and of blinding headlights at night. In addition, plants significantly reduce harmful gaseous and particulate pollutants in the atmosphere, while effectively reducing noise pollution. Finally, they provide an aesthetic buffer of beauty and softness against the harsh edges and hard lines of urban environments.

Plants as Separators

Trees along the street act as subtle but effective physical and psychological separators between pedestrians and automobiles. As spatial separators, they create transition zones between the activity of the street and the less active spaces adjacent to buildings.

Automobiles are separated from pedestrian traffic by combinations of tree and shrub plantings. In urban areas, large cement or terra cotta planters and sidewalk cutouts are used, while in residential areas often there is a continuous planting strip between street and sidewalk. It would seem ideal to provide this same kind of planted separation for bicycles when bikeways run adjacent to automobile traffic.

FIGURE 24

Roadway Planting Separators

Trees along the roadway separate pedestrians and traffic.

Unfortunately, if plants were used as separators between automobiles and bicyclists, and similarly between pedestrians and bicyclists, an already overcrowded and complex traffic profile would become even more confusing. Furthermore, this practice would require greater amounts of valuable land for transit. However, high-use arterials or grand boulevards might warrant this spaciousness and protection, and this type of planting could be pleasantly effective.

FIGURE 25

Plantings for Transition between Spaces

Trees smooth the transition from traffic to adjacent spaces.

Less urbanized areas, such as planned developments and recreation areas, present much greater opportunities for the use of plants. In these areas, it is often possible to use Class I bikeways, which by their nature involve greater physical separation from other transit modes. They are often included in parks, greenbelts, planned unit developments, and college campuses, and may be used in a number of new corridor possibilities now being invicstigated for bikeway use. Pedestrian paths often lie immediately adjacent or

parallel to Class I bikeways, and for this reason it is often necessary
to plan both bikeway and path as a single unit. The close proximity
of bicycles and pedestrians requires some form of separation or de-
marcation.

The University of California at Davis has experimented with
the separation problem between bicyclists and pedestrians and has
devised a solution using only pavement changes and trees. The bikeway is asphalt and the pedestrian walk is made of decomposed
granite rolled smooth. Trees are planted approximately 20 feet
apart in small square cutouts between the two surfaces. Except for
the trees, there is no separation of traffic other than the pavement
change, allowing easy cross-access by pedestrians or bicyclists. The
single row of trees provides a pleasant overhead canopy, giving the
space a more human scale and a comfortable sense of security.

FIGURE 26

Plantings for Traffic Separation

Plantings can be used as separators between three different
transit modes, where traffic is heavy and adequate space is available.

FIGURE 27

Paving Materials for Delineation

Decomposed granite and asphalt delineate spaces for pedestrians and bicyclists at the University of California at Davis.

Plants as Screens

The use of plant materials for screening as well as separation may also be necessary, because often the night lights of automobiles temporarily blind bicyclists, particularly at intersections. Where bicyclists ride adjacent to the street, a simple hedge or mass planting can effectively screen automobile headlights. This same screen also serves to block out drivers' views of bicycle lights which at times can be confusing. Planting the outer curve of a two-way bikeway before it crosses a street can also prevent the temporary blinding of bicyclists at this critical point.[45]

FIGURE 28

Roadway Planting Screens

Plantings can be used to screen automobile headlights.

FIGURE 29

Roadway Intersection Screening

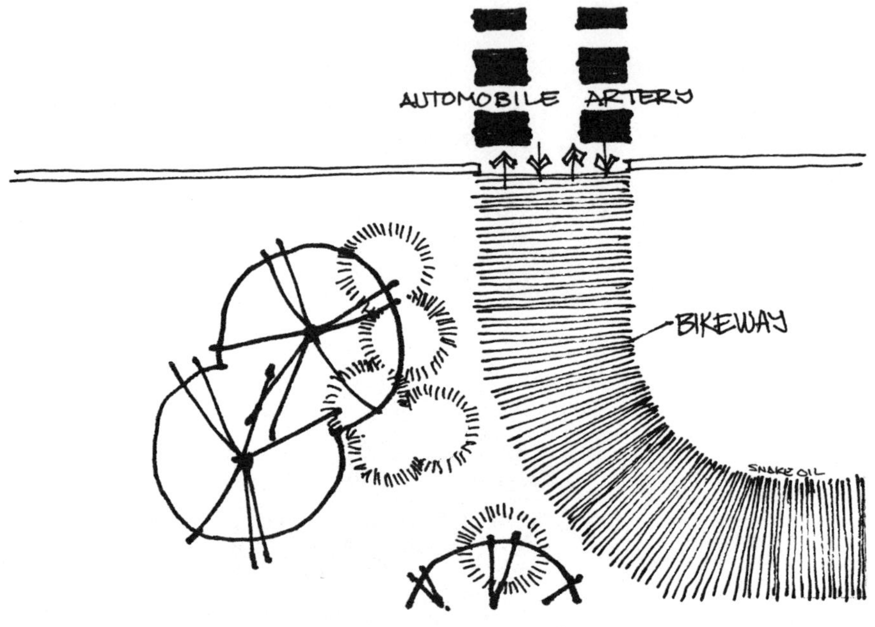

Screening is critical at intersections where bicyclists can be temporarily blinded. (*Fietspaden en -oversteekplaatsen,* Nederland)

Plants and Climate Control

When people leave the highly protected environment of architectural space, plant materials take over the job of minimizing and softening the outside elements. Plants have long been used to control wind by obstruction, guidance, deflection, and filtration. Deciduous trees provide shade in summer and allow the warming sun

TABLE 1

Plant Materials for Screens

Name	Evergreen(E) or Deciduous(D)	Height	Geographical Region	Remarks
Berberis thunbergii Japanese Barberry	D	6 ft.	Midwest, Northwest	Dense, thorny vigorous growth; bright red fall fruit
Eucalyptus ficifolia Red Flowering Gum	E	40 ft.	West coastal areas only	Large clusters flowers, pink, coral, red
Fagus sylvatica European Beech	D	90 ft.	Northwest, East	Dense foliage; large canopy
Gleditsia triacanthos "Inermis" Thornless Common Honeylocust	D	75 ft.	Northwest, East	Widespread canopy
Euonymus japonica Evergreen Euonymus	E	10 ft.	South, West	Compact; excellent foliage
Liquidambar styraciflua American Sweet Gum	D	60 ft.	West, East	Dense cover; excellent fall color
Magnolia grandiflora Southern Magnolia	E	90 ft.	West, South, Northwest	Dense cover; large white spring early summer flower
Viburnum	E or D	8-10 ft.	West	Striking spring flowers, fruit in summer
Pinus contorta Shore Pine	E	30 ft.	West, East	Dense foliage
Pinus mugo Swiss Mountain Pine	E	35 ft.	Midwest, Northwest	Dense and globose growth
Platanus acerifolia Sycamore	D	50 ft.	West, East	Hardy for street conditions; handsome bark
Sorbus aucuparia European Mountain Ash	D	70 ft.	Northwest, East	Bright red fruit in early fall

Source: Alfred C. Hottes, *The Book of Trees* and Donald Wyman, *Wyman's Gardening Encyclopedia.*

to penetrate in winter. Coniferous evergreens can absorb as much as 40 percent of the rain which falls on them. Trees, shrubs, and ground covers also effectively minimize extreme day-to-night temperature changes.

Plants and Rain

Keeping dry on a bicycle in wet weather can be a real problem. Protective clothing and bicycle fenders can keep bicyclists dry, but

FIGURE 30

Rain Wear for Bicycling

Rain ponchos provide some protection for bicyclists in bad weather.

in doing so they often restrict freedom of movement and effective release of body heat while pedaling, and often there is an incongruity between rain gear and the attire required at the bicyclist's ultimate destination. A canopy of coniferous and deciduous trees over a bikeway can help deal with the problem by reducing the amount of rainfall which reaches the bicyclist by as much as 40 and 20 percent, respectively.[46]

The amount of rain which reaches the ground through the trees is dependent upon the intensity and duration of the rainfall, type of trees, and the structure of the tree canopy. "Some studies

FIGURE 31

Plantings for Rain Protection

Tree canopies afford bicyclists protection from the rain.

have indicated that only 60 percent of the rain falling on a pine forest [conifer] reaches the ground, and 80 percent of the rain falling on a hardwood [deciduous] forest canopy reaches the ground."[47] "This might seem surprising, since the [deciduous] foliage is denser than it is in a pine forest, but while the rain remains in drops on the pine needles, it runs down the broad leaves of deciduous trees, and flows down the branches. This happens so effectively that the amount of rain to reach the ground may exceed 50 percent, even in very light rain. The water running down the trunk amounts to one-fifth of the total, whereas less than 5 percent runs down the trunk of the pine."[48]

In rainy regions, trees planted along bikeways can help a great deal in protecting bicyclists from the weather. All trees provide some protection against rain, and certain species provide greater protection than others; designers should keep this in mind while selecting tree types for bikeways. However small the improvement, bicyclists will be intimately aware of it.

Plants and Wind

Wind can also be a major difficulty for bicyclists, especially in areas with exceptionally strong prevailing winds. Riding with the wind can be a boon to bicyclists, but if they have to return against the wind they may be forced to walk their bikes. Wind also causes problems by stirring up street dust, channeling polluted air through critical areas like parks and other high-bicycle-use areas. Plants can be used effectively to increase or decrease wind speed and to change the direction of airflow. The degree to which plants are effective depends upon their physical structure and upon their arrangement.[49] A dense grove of trees will cut to a minimum the speed of wind traveling under them. A row of trees called a *shelter belt,* running perpendicular to the wind, can reduce wind speed by 50 percent for a distance downwind of from ten to twenty times the windbreak height.[50] It is interesting that a slightly more penetrable windbreak will provide a wind speed reduction for a greater distance downwind. The air which flows through helps to fill the vacuum created downwind of the windbreak. As far as changing

TABLE 2

Plant Materials for Rain Control

Name	Evergreen(E) or Deciduous(D)	Height	Geographical Region	Remarks
Acer platanoides Norway Maple	D	90 ft.	Northwest, East	Densely branched round head, excellent foliage
Catalpa speciosa Northern Catalpa	D	70 ft.	Northwest, East	Tropical effect, broad canopy
Crataegus lavallei Lavalle Hawthorn	D	25 ft.	West, East	White flower in spring, very large orange fruit
Eucalyptus ficifolia Red Flowering Gum	E	40 ft.	West coastal areas	1 ft. flower clusters, pink, coral, red
Fagus sylvatica European Beech	D	90 ft.	Northwest, East	Extremely large canopy, dense foliage
Juniperus virginiana Eastern Red Cedar	E	70 ft.	West, East	Dense and hardy
Pinus contorta Shore Pine	E	30 ft.	West, East	Dense foliage
Pinus nigra Austrian Black Pine	E	90 ft.	West, East	Windbreak, dense
Prunus lusitanica Portugal Laurel	E	60 ft.	West, East	Dense, tolerates wind, and hot sun
Quercus palustris Pin Oak	D	75 ft.	Northwest, Midwest, East	Dense foliage, excellent fall color
Quercus robur English Oak	D	90 ft.	East, West	Dense foliage, excellent canopy
Tilia cordata Little-Leaf Linden	D	50 ft.	Northwest, East	White flower, late spring, early summer

Source: Alfred C. Hottes, *The Book of Trees* and Donald Wyman, *Wyman's Gardening Encyclopedia.*

FIGURE 32

Plantings as Shelter Belts

Plantings used as shelter belts can reduce wind speed significantly.

wind direction, trees, shrubs, or small groupings of plants placed individually can be used to direct airflow in specific directions or to increase circulation necessary in keeping polluted air out of critical areas.

Plants and Temperature

Heat and cold are two important elements in personal comfort. Automobiles have heaters, air conditioners, and vents which regulate air temperature for passenger comfort, but pedestrians and bicyclists must regulate their temperature by modifying their clothing. They have no control over air temperature and humidity or the direct and reflected rays of the sun. However, the

TABLE 3

Plant Materials for Windbreaks

Name	Evergreen(E) or Deciduous(D)	Height	Geographical Region	Remarks
Berberis thunbergii Japanese Barberry	E	6 ft.	Midwest Northwest	Dense, thorny vigorous growth; bright red fall fruit
Gleditsia triacanthos "Inermis" Thornless Common Honeylocust	D	75 ft.	Northwest, East	Widespread canopy
Euonymus japonica Evergreen Euonymus	E	10 ft.	South, West	Compact; excellent foliage
Juniperus virginiana Eastern Red Cedar	E	70 ft.	West, East	Dense and hardy
Picea pungens Colorado Spruce	E	80 ft.	West, East	Extremely dense, pyramidal tree
Pinus mugo Swiss Mountain Pine	E	35 ft.	Midwest, Northwest	Dense and globose growth
Populus tremuloides Quaking Aspen	D	50 ft.	Northwest, West (alpine conditions)	Vigorous growth brilliant yellow fall color
Quercus macrocarpa Bur oak	D	75 ft.	Northeast	Dense cover
Syringa vulgaris Lilac	D	20 ft.	Northwest, Northeast	Fragrant spring flower
Taxus cuspidata Japanese Yew	E	50 ft.	Midwest, East	Compact growth; dark green foliage
Thujia occidentalis American Arborvitae	E	60 ft.	Northeast, Northwest	Upright open growth
Tsuga canadensis Canada Hemlock	E	90 ft.	East, Northwest	Dense, pyramidal tree

Source: Alfred C. Hottes, *The Book of Trees* and Donald Wyman, *Wyman's Gardening Encyclopedia.*

FIGURE 33

Plantings for Temperature Control

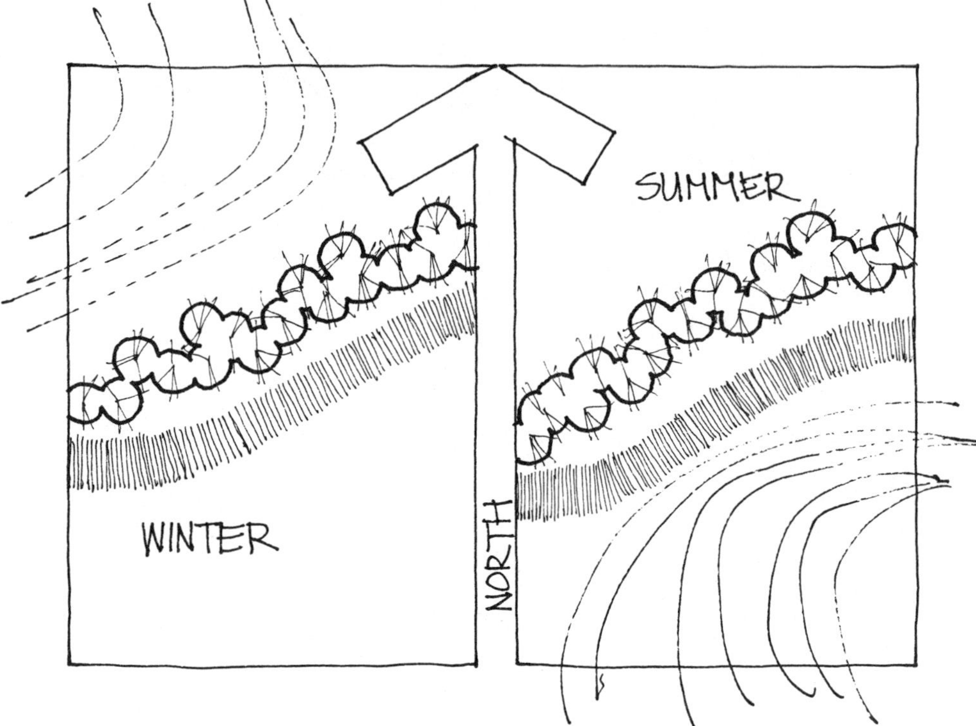

Plants can be used to block harsh winter winds and to collect cooling summer breezes.

transportation planner does have considerable control over these elements, and his most effective tools are plant materials.

Trees, shrubs, groundcovers, and vines all play extremely important roles in regulating bikeway and pedestrian path temperatures. For example, deciduous trees provide shade in summer and allow the sun's warming rays to penetrate to the ground in the winter. They also reduce air temperature beneath them:

FIGURE 34

Deciduous Trees for Temperature Control

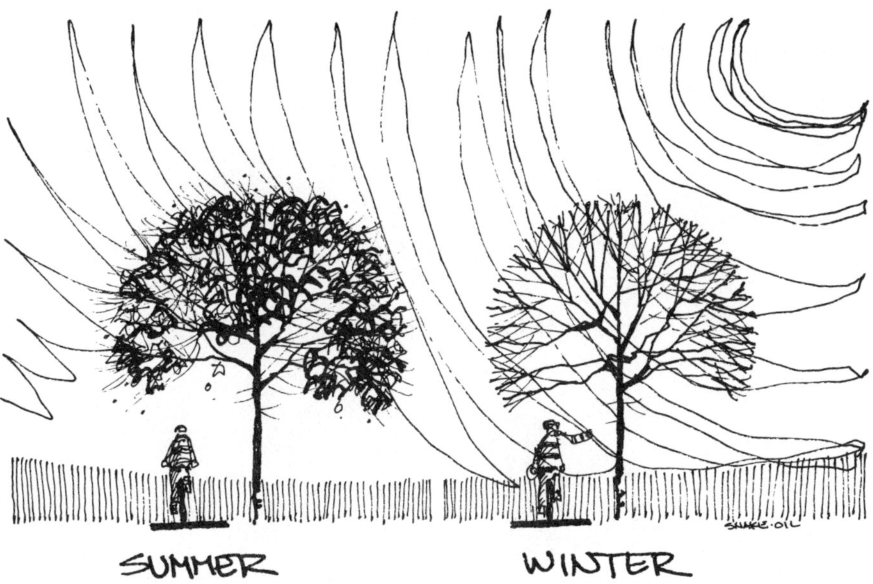

Deciduous trees provide cool shade in summer and let the warming sun through in winter.

The air layer beneath the trees receives some chilled air from the foliage canopy overhead, but the main source of cool air would be from a moist turf [or ground cover], if present. In hot-dry climates shading and moisture-producing surfaces have a very great cooling effect on the air unless the wind is strong. In hot-humid climates air circulation is necessary to encourage evaporation and large, tall trees with an open lawn [or ground cover] are desirable.[51]

A tree canopy can also minimize diurnal temperature changes by trapping heat which normally would have radiated from the ground into space. The canopy's ability to hold warm air is directly related to the density of the foliage. "The rate of heat loss by reradiation, retarded by plant cover, results in nocturnal temperature variations of both the soil under and the air within the plant cover, with temperatures not dropping as low as the soil and air temperature of adjacent areas."[52]

FIGURE 35

Plantings for Containment of Daytime Heat

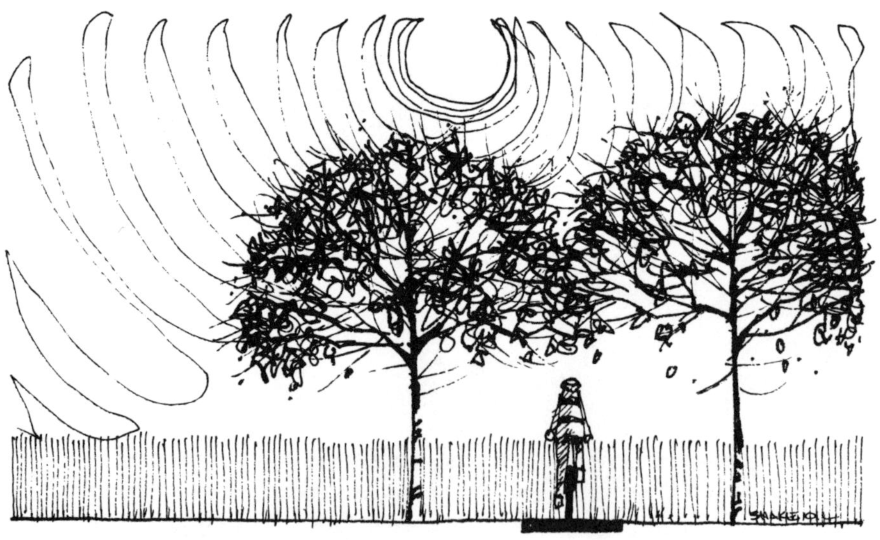

At night, tree canopies trap the day's heat, which otherwise would radiate into space.

Plants and Snow

Plant materials can be used to control the drifting of snow and the speed at which it melts; this control is critical to nearly all forms of transportation in regions where it snows regularly. Wind needs a certain velocity to pick up and carry snow; if the wind is slowed below this velocity, it drops the snow. A snow fence reduces wind velocity and causes the snow to fall before it reaches the road, but the fence requires expensive installation and yearly maintenance. A windbreak of trees and shrubs, on the other hand, requires little or no care once established, can be much more effective than a fence, and is visually pleasing. Trees can also influence the speed at which snow melts from a road or bikeway: if the surface is shaded, snow and ice will remain much longer. Therefore, the placement, size, and variety of trees adjacent to a road or bikeway are all important in snow regulation.

Plants and Air Pollution

As mentioned earlier, vegetation alongside roadways can help to cleanse the air of industrial dust particulates, a major form of atmospheric pollutant currently found in the urban environment. Gaseous pollutants comprise the other major form contributing to today's urban atmospheric pollution, and they account for the most widespread injury to plant life. "Gases known to damage vegetation include ozone, PAN (peroxyacetylnitrate), nitrogen dioxide, sulfur dioxide, hydrogen fluoride, ethylene, and chlorine. These pollutants destroy plant chlorophyll, disrupt the photosynthesis process, and consequently reduce food production."[53] Only a few of these gases, such as ozone (O_3), peroxyacetylnitrate (PAN), ethylene ($CH_2=CH_2$), and oxides of nitrogen (NO_X) are directly caused by automobile emissions. The others are produced by industrial manufacturing and the generation of electricity. All of these gaseous pollutants can be a real health hazard to individuals engaged in recreational activity.

TABLE 4

Plant Materials for Snow Control

Name	Evergreen(E) or Deciduous(D)	Height	Geographical Region	Remarks
Abies concolor White Fir	E	120 ft.	Midwest, North, East, Northwest	Dense cover resistant to heat and drought
Abies grandis Grand Fir	E	120 ft.	West coast	Cool moist climate, dense cover
Juniperus virginiana Eastern Red Cedar	E	70 ft.	West, East	Dense and hardy
Picea pungens Colorado Spruce	E	80 ft.	Midwest, West	Extremely dense, pyramidal tree
Pinus contorta Shore Pine	E	30 ft.	West coast, East	Dense foliage
Pinus nigra Austrian Black Pine	E	90 ft.	West, East	Windbreak; dense
Pinus thunbergii Japanese Black Pine	E	90 ft.	Northeast, Northwest	Growth irregular spreading with age
Quercus macrocarpa Bur Oak	D	75 ft.	Northeast	Dense cover
Quercus rubra Red Oak	D	75 ft.	East, Northwest	Excellent canopy dense foliage
Quercus palustris Pin Oak	D	75 ft.	Northwest, Midwest, East	Dense foliage excellent fall color
Thujia plicata Giant Arborvitae	E	180 ft.	Northwest, Western coast regions	Dense cover windbreak
Tsuga canadensis Canada Hemlock	E	90 ft.	East, Northwest	Dense, pyramidal tree

Source: Alfred C. Hottes, *The Book of Trees* and Donald Wyman, *Wyman's Gardening Encyclopedia.*

Convincing evidence points to the deleterious effects of air
pollution on the health of general populations. Increased
morbidity and mortality rates from lung cancer, stomach
cancer, emphysema, and other respiratory diseases have
been noted for populations in highly polluted areas. Persons
engaging in recreational activities near major concentrations
of automobile traffic are probably subjecting themselves
to particularly high levels of air pollution. The more vigor-
ous the recreational activity the more hazardous may be
the effect of the pollution. It is important that planners
take these factors into account when making recreational
land use decisions.[54]

Levels of air pollution along major arterials may be
as much as ten times higher than levels of pollution in general community ambient air. . . .[55]

. . . with heavy breathing, particulates may be pulled
deeper into the lungs and deposited. Also, a greater volume
of gaseous pollutants may be pulled over the particulates,
increasing the possibility of synergistic reactions. Lead is
another toxin which may enter the body faster with exercise.[56]

Thus automobile emission pollutants can directly affect the
bicyclist during heavy respiration and may be decreased either by
reducing pollution near bikeways or locating bikeways in unpol-
luted areas. Since the latter solution is not always possible, we must
find methods for reducing pollution in bikeway corridors, especially
those containing heavy automobile traffic. There are a number of
ways to do this: enact strict emission control legislation for both
automobiles and industry, provide good air circulation, and establish a street tree program which can help cleanse the air within the
corridors. Though gaseous pollutants from automobiles and
industry should be reduced or eliminated and air should always be
kept from stagnating, planners have little control over this aspect
of the pollution problem. However, some control may be exercised
by planners in the selection of plantings which are resistant to
atmospheric pollutants and can reduce their effect.

In addition to channeling polluted air elsewhere, street plantings can help to reduce dust particles. Research indicates that there

TABLE 5

Plant Materials for Pollution Control

Name	Evergreen(E) or Deciduous(D)	Height	Geographical Region	Remarks
Acer platanoides Norway Maple	D	90 ft.	Northwest, East	Densely branched round head, excellent foliage
Acer saccharum Sugar Maple	D	120 ft.	Northwest, Northeast	Pendulous, arching branches; excellent fall color
Betula pendula European Birch	D	60 ft.	Northwest, Northeast	Pyramidal form, short-lived tree
Cornus racemosa Gray Dogwood	D	15 ft.	East	Numerous small white flowers in late spring
Picea abies Norway Spruce	E	150 ft.	Northwest, East	Fast-growing, drooping branches
Picea glauca White Spruce	E	80 ft.	Northeast	Dense when young
Picea pungens Colorado Spruce	E	80 ft.	West, East	Extremely dense, pyramidal tree
Quercus robur "Fastigiata" Pyramidal English Oak	D	75 ft.	West, East	Dense foliage, columnar form
Quercus rubra Red Oak	D	75 ft.	East, Northwest	Excellent canopy, dense foliage
Thujia occidentalis American Arborvitae	E	60 ft.	Northwest, Northeast	Upright, open growth
Ulmus parvifolia Chinese Elm	D	50 ft.	Northwest, East	Exfoliating bark, excellent fall color

Source: Donald D. Davis, *Air Pollution Damages Trees* and Ibrahim Joseph Hindawi, *Air Pollution Injury to Vegetation.*

are 3,000 kernels per air units in streets planted with trees, as compared to between 10,000 and 12,000 particles in streets without tree plantings.[57] The leaves absorb and detoxify many millions of tons of air pollutants, and the release of oxygen from trees is believed to lower the carbon monoxide levels.[58] Leaves can further act as filters, collecting dust which rain later washes into the soil and adds to the humus layer. Thus it is clear that street plantings can serve a functional as well as aesthetic purpose in transportation design.

ENGINEERING

Engineers find themselves in little agreement over the working relationship between the bicycle, the automobile, and the pedestrian. The main reason is increasing variation in bikeway standardization processes. Some engineers contend that bikeways should be accessible to pedestrians at certain times, while others feel this would be dangerous for pedestrians and would further obstruct the free flow of traffic. This is merely one indication that some measure of standardization must occur before questions such as bikeway width, removal of on-street parking, uniformity of signs, and intersection design can be properly handled.

Standardization

System standardization gives the bicyclist a sense of confidence when traveling throughout the area. If the standardization is extended to the regional level, both visiting and local motorists can travel safely through any part of the system without difficulty or apprehension concerning pedestrian and bicycle traffic; and a child who learns to bicycle in his own neighborhood can safely travel through other parts of the system. Inter-regional coordination of engineering standards could eventually facilitate the creation of national bikeway standards. It should be remembered, however, that the attempt to implement standardized engineering procedures should not interfere with the solution of a specifically local problem. Specific problems can usually be handled much more imaginatively and effectively by someone familiar with engineering concepts appropriate for the local situation.

There are a number of reasons why standardized classifications for bikeway systems should be established. Consistency within the bikeway system allows traffic engineers to provide safety, convenience, and comfort through routes of differing purposes. Standard signs would inform bicyclists of hazardous situations and motorists of their obligations, by identifying specific classes of bikeways. In addition, these signs would inform pedestrians of their rights and of the restrictions regarding different types of bikeways.

Engineers need a means of graphically designating the various classes of bikeways. Graphic signs could indicate the safety gradients of the different types of bikeway; for example, motorists driving on designated freeways know they will not encounter cross-traffic, trains, or bicycles, and on roads not labeled as freeways they will. Pedestrians safely assume they will not find automobiles on sidewalks. Therefore, bicyclists should not expect automobiles or pedestrians on Class I bikeways which have been so designated. In cases where this might occur, they should be forewarned with signs. They should also be cautioned to expect frequent intersections when that is the case. When the bikeway is free of intersections and other traffic modes, bicyclists should be informed that they are traveling on a route similar to a freeway.

If uniform classifications are to be set up, bikeways should be classified into as few groups as possible. Most bikeway planning literature uses three classifications similar to the ones described earlier in the chapter, with each class denoting varying degrees of safety, comfort, and convenience (safety being the most critical).

Safety, comfort, and convenience factors revolve mostly around bicyclists' conflicts with automobiles and pedestrians. A Santa Barbara, California, study found that "82 percent of collisions are related to maneuvers by the bicyclist to enter into, to turn across, or to cross through a flow of traffic."[59] Therefore, in Class I bikeways, crossings of automobile traffic must be minimized or avoided through the use of over- or underpasses.

A more controversial question is whether Class I and II bikeways should allow for shared bicycle and pedestrian traffic.

FIGURE 36

Bikeway Undercrossing

Bikeway undercrossing, University of California at Davis.

Accidents involving a bicyclist and a pedestrian are rarely fatal and
are almost never reported. None-the-less, they are a frequent problem and a common annoyance to both parties, and deserve attention. Since Class I bikeways often serve recreational purposes, it is
vital that annoyances and apprehensions about collision or injury be
eliminated. Hikers and pedestrians should be given defined and adequate space within a bikeway when separate facilities are not
spatially or economically feasible; a separate path is the most
preferable solution.

FIGURE 37

Sidewalk Shared by Pedestrians and Bicyclists

Conflicts between pedestrians and bicyclists are common on shared sidewalks.

The same arguments apply to Class II bikeways which use sidewalk space. Pedestrians on sidewalks do not expect to find bicyclists there, and average sidewalks are rarely wide enough to allow bicyclists to pass two pedestrians walking side-by-side. The problem is further aggravated because pedestrians do not normally conform to vehicle rules, such as keeping to the right and signaling turns or abrupt movements. Bicyclists on sidewalks are often forced to a complete stop before pedestrians realize their presence.

Physical Requirements of Bikeways

Generally there has been one approach to specifying such requirements as bikeway width, vertical clearance, and radius of curvature and grades. Minimum standards have been based upon the absolute physical limitations of the bicycle and the average bicyclist. While absolute minimums are a valid and necessary part of any standardization program, it is often difficult to obtain information on optimum conditions for the bicyclist. It is clear that comfort, convenience, and safety become greater as widths, vertical clearances, and the radii of curvature increase, and as steepness of grades decreases. There is, however, a point at which these increases provide little added benefit, and when this occurs, comfort, convenience, and safety are at a relatively high level. German specifications for a one-way, two-lane bikeway set the absolute minimum paving width at 1.7 meters, or 5.3 feet;[60] the State of California recommends a comfortable maneuvering allowance of 6.4 feet because a width greater than that provides little added benefit.[61] In an attempt to deal with this problem, the following section presents both accepted absolute minimums and comfort allowances.

Selection of Bikeway Class

When the need for a specific bikeway between a residential and a commercial area has been determined, corridor selection is the first consideration. In choosing proper corridors, a brief investigation of the particular situation and available path locations should be made.

For example, perhaps it has been established that there is a need for a bikeway between a suburban school district and a more commercialized shopping area. The available corridors include a major automobile arterial, a number of lightly traveled side roads, and an abandoned railroad right-of-way. Due to its unique separation, the railroad right-of-way is likely to be the only location in which a Class I bikeway could be established. The side roads are wide enough to accommodate a Class II or III bikeway, while the

FIGURE 38

Class I Bikeway Width

Class I bikeway width: a width of 6.5 feet allows space for two-way traffic; a width of 10.9 feet allows enough room for one bicyclist to pass two others riding side-by-side. (*Comprehensive Bikeway Plan*, City of Seattle, Washington; *Bikeway Planning Criteria and Guidelines*, State of California)

FIGURE 39

Class II Bikeway Width, One Lane

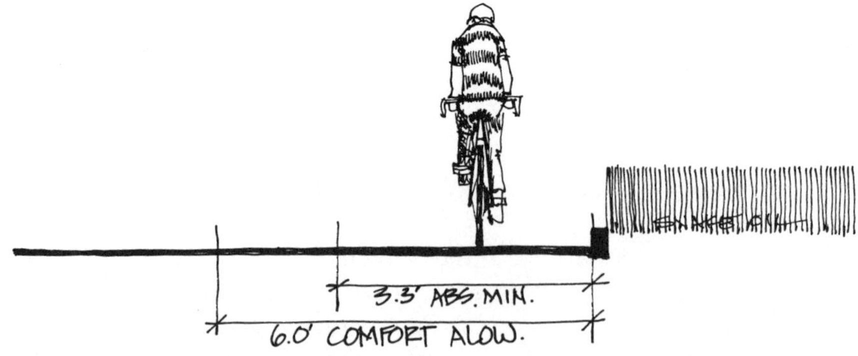

Class II bikeway width, one lane: the German minimum for one bicyclist is 1 meter, or 3.3 feet; a width of 6 feet allows room for three-wheeled pedaled vehicles and wheelchairs. (*Radverkehrsanlagen Richtlinien*, Deutschland; *Bikeway Design*, State of Oregon)

FIGURE 40

Class II Bikeway Width, Two Lanes

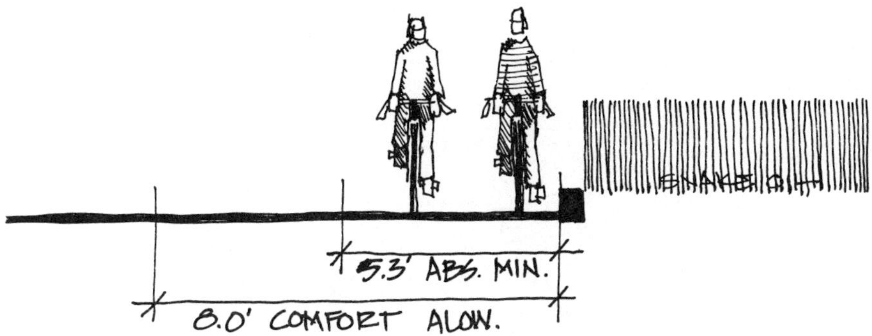

Class II bikeway width, two lanes: the German minimum of 1.7 meters, or 5.3 feet, provides space for bicyclists to pass one another; an 8-foot width allows comfortable side-by-side riding. (*Radverkehrsanlagen Richtlinien,* Deutschland; *Bikeway Design,* State of Oregon)

FIGURE 41

Vertical Clearance

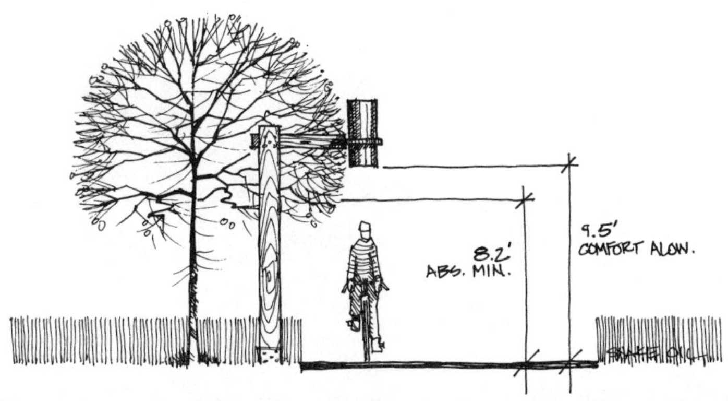

Bikeway vertical clearance: German specifications give 2.5 meters, or 8.2 feet, as the absolute minimum, while Oregon suggests a more comfortable 9.5 feet. (*Radverkehrsanlagen Richtlinien,* Deutschland; *Bikeway Design,* State of Oregon)

FIGURE 42

Bicycle Grade-Length of Grade Relationships

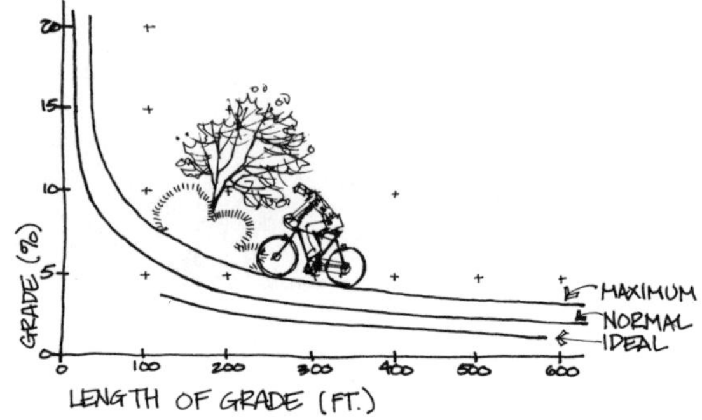

Relationship between degree of grade and length of grade: many planning agencies use these Dutch standards, while setting a general limit of 5 percent. (*Bicycle Circulation and Safety Study,* City of Davis, University of California; *BART/Trails,* Oakland, California; *The Bicycle,* Atlanta Metropolitan Region)

FIGURE 43

Radius of Curvature and Velocity Relationship

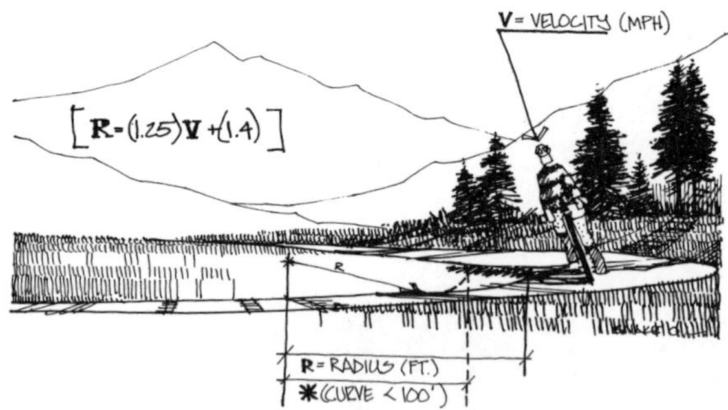

Relationship between radius of curvature and velocity: V = velocity, or miles per hour; R = radius of curvature, in feet; and R = 1.25 V + 1.4. The required radius can be computed for various bikeway conditions, such as steep downgrades. Curves with radii of less than 100 feet should be widened on two-way bikeways, but not by more than 4 feet; see asterisk. (*Bikeway Planning Criteria and Guidelines,* State of California; *Bikeway Design,* State of Oregon)

major arterial has the capacity to support only Class III bikeways. If the engineer considers convenience, economics, politics, land use, and safety, he logically chooses the railroad right-of-way for the bikeway. However, the land may be more difficult to obtain because of high cost. In view of the fact that the major arterial provides the least safe solution, utilization of the lightly traveled side roads for a Class II bikeway then becomes the best solution. In the event that funds are not immediately available for construction of the Class II system, the side roads could be temporarily designated as a Class III bikeway. If on-street parking or mitigating circumstances prohibit use of the side roads for either Class II or III bikeways, the major arterial could be temporarily classified as a Class III bikeway, but this would only be a last resort..

In general, when selecting a corridor for a bikeway, the engineer should locate all possible corridors and evaluate them in terms of possible bikeway class, safety, bicyclist convenience, access and directness of route, time required from planning to construction, land use, political implications, and cost. While planning a bikeway system, he should keep interim facilities in mind.

Intersection Design

Planning for crossings and intersections is one of the most serious problems the bikeway designer faces. Intersections with only two traffic modes—pedestrian and automobile—are dangerous in themselves. The addition of bicycles without provisions for their recognition or safety can be disastrous. The Class I bikeway ideally eliminates all automobile and pedestrian cross-traffic. Class III bikeways provide no special facilities for bicyclists other than warning signs. The Class II bikeway, however, must provide for the bicyclist at intersections as well as along the road. Too often, bikeways simply disappear at intersections, reappearing and continuing on the other side. In these situations, the most dangerous condition along the bikeway has been ignored.

Studies in the Netherlands are an excellent resource on this subject (see Bibliography), and many of the more successful solutions in the United States have been based on this work. The

FIGURE 44

Class I Bikeway

Class I bikeways are recommended for high-use bicycling, either for transportation or recreation.

FIGURE 45

Class II Bikeway, Above Curb

Class II bikeways should be above the curb when near heavily-traveled or high-speed roads.

FIGURE 46

Class II Bikeway, Below Curb

Class II bikeways placed below the curb are recommended when the bikeways are adjacent to low-use or low-speed roads.

FIGURE 47

Class II Bikeway, Below Curb, Parking Prohibited

Class II bikeways below the curb, with parking prohibited, are recommended if the bikeways are adjacent to motorized traffic.

FIGURE 48

Class III Bikeway, Shared Pedestrian and Bicycle Path

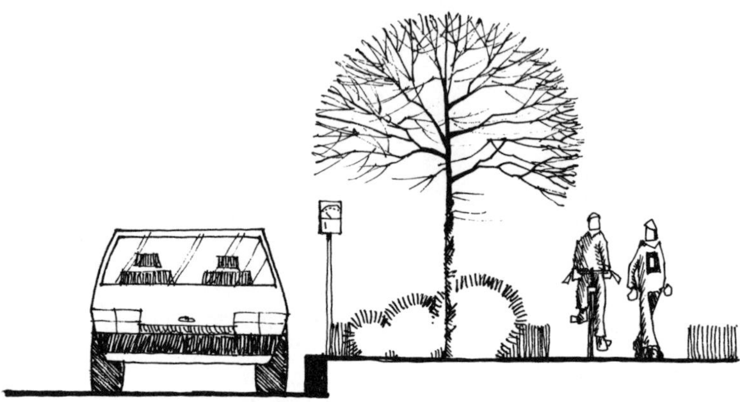

Class III bikeways in which pedestrians share the path are recommended only for temporary purposes but are safer than shared bicycle and automobile routes.

FIGURE 49

Class III Bikeway, Shared Bicycle and Automobile Route

Class III bikeways which share the route with automobiles are recommended only for temporary purposes if width, shoulder, traffic intensity, and traffic speed allow.

FIGURE 50

Intersection Design for Class I Bikeways

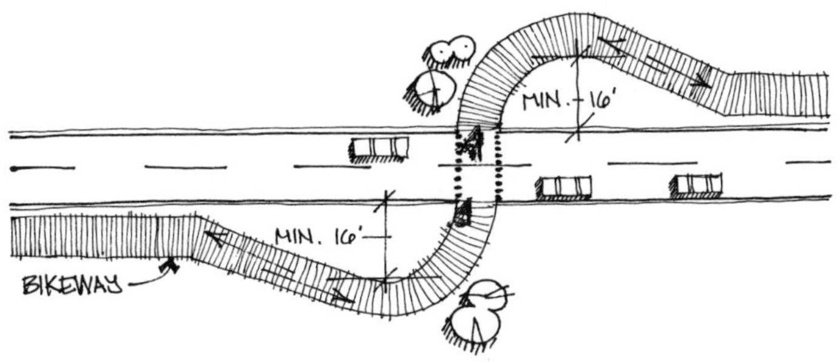

Intersection design for Class I bikeways: when an over- or underpass is not possible, the bikeway should cross automobile traffic at a right angle so that bicyclists can see traffic driving in the same direction. (*Fietspaden en -oversteekplaatsen*, Nederland)

FIGURE 51

Intersection Design for Class II Bikeways

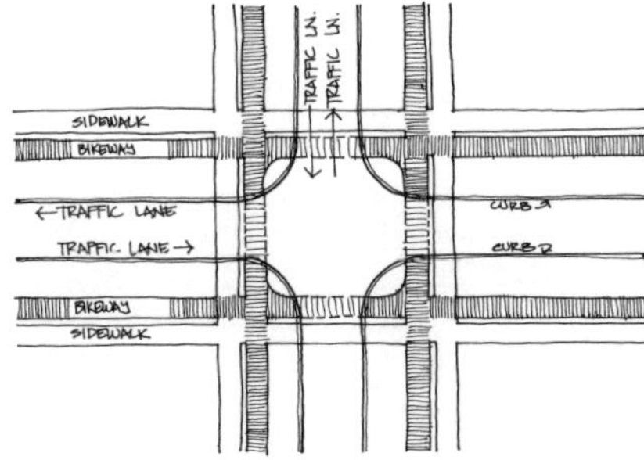

Intersection design for Class II bikeways above the curb: little modification of pedestrian and automobile movement is needed, and this system might be most valuable at intersections with heavy automobile, bicycle, and pedestrian traffic. Bicyclists traveling straight through are not led out of direction, and left-turning bicyclists make one simple, 90-degree turn but must cross through two signal phases. Areas are provided for bicyclists to wait safely while signals change or traffic clears, so that they will not obstruct pedestrian, bicycle, or automobile cross-traffic.

FIGURE 52

Intersection Design for Class II Bikeways, Below Curb

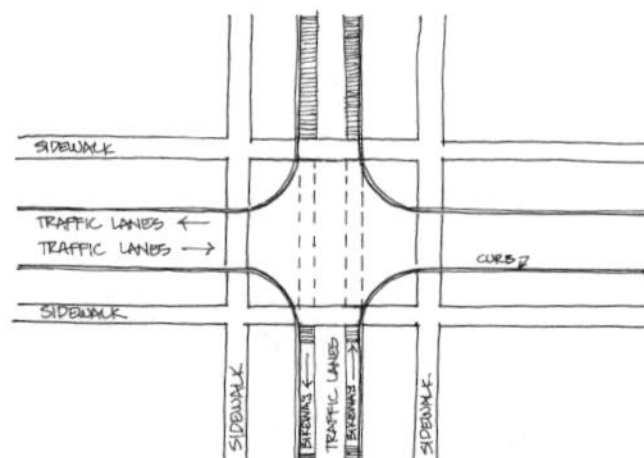

Intersection design for Class II bikeways below the curb: this is perhaps the simplest and most effective design for light to moderately traveled roads. Straight-through bicyclists are not led out of direction, and left-turning bicyclists may move easily to the center lane to turn with other traffic. The dashed crossing lines for the bicycle are important because they warn cross-traffic that bicyclists should be expected.

FIGURE 53

Intersection Design for Class III Bikeways, Pedestrian and Bicycle Path

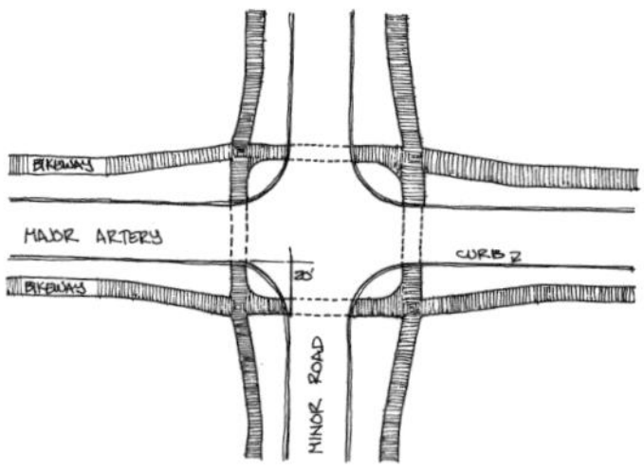

Intersection design for Class III bikeways which are shared with pedestrians: the paths cross the roads approximately 20 feet from cross-traffic, and this enables straight-through bicyclists and right-turning automobiles to cross at right angles. This insures that the automobile will be in the bicyclist's forward field of vision. Waiting areas are provided for bicyclists and the 20-foot setback allows space for one automobile to wait between bikeway and cross-traffic. This solution may be valuable for moderate to heavy recreational bicycling with light pedestrian use, because of the generous space provided. (*Fietspaden en -oversteekplaatsen*, Nederland)

Dutch treat the bicycle as a completely separate form of transportation; they provide marked crossings and separate signals and signs in the same way that we provide crosswalks and signals for pedestrians.

Because intersections are the most dangerous points along any given bikeway, it is absolutely necessary that they be of prime concern to the designer/engineer. If intersections along a particular route present problems too difficult to handle properly, it would be best to delay construction until a proper solution is found or to relocate the bikeway so that the intersection is avoided.

Barriers and Separators

FIGURE 54

Planted Barrier between Bicyclists and Pedestrians

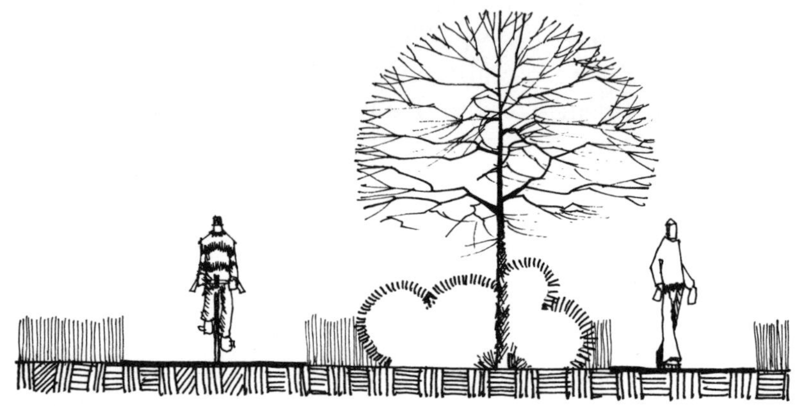

The need for additional space is one drawback to using a planted barrier between bicyclists and pedestrians.

FIGURE 55

Planted Barrier between Bicyclists and Motorists

A planted barrier between bicyclists and motorists also has the drawback of taking up additional space, and furthermore, confusion can result at intersections where the planted island must be broken.

FIGURE 56

Extruded Curb Barrier

Advantages of an extruded curb barrier are low cost and minimal space requirements, but it creates confusion where it must be broken. An additional problem is that automobiles cannot cross this barrier to park at the curb or enter driveways.

FIGURE 57

Painted Line Barrier

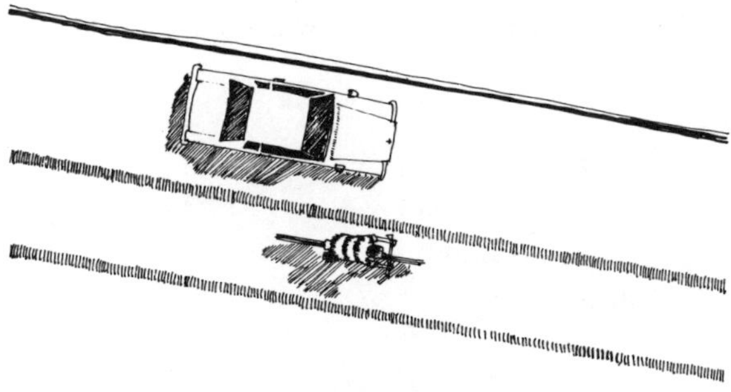

A painted line barrier is perhaps the simplest and least expensive separator. Painted lines can continue through intersections and allow bicyclists to leave their lane and merge into the center traffic lane to make left turns. Painted lines also allow automobiles to cross through the bicycle lane in order to park or enter driveways. Because this barrier will not stop automobiles from entering the bikeway it should not be used on high-speed roads.

FIGURE 58

Full-width Colored Pavement Barrier

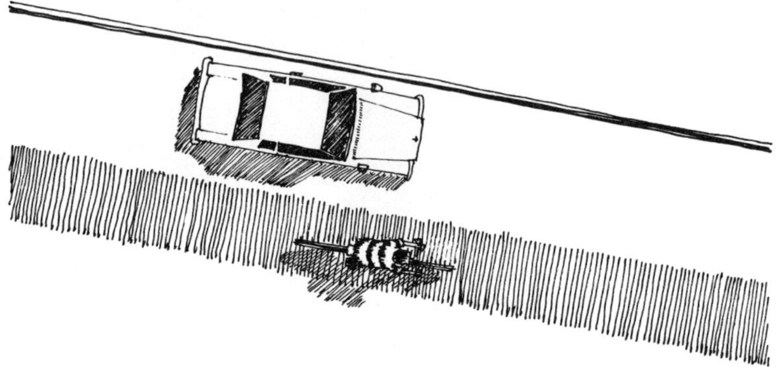

Pigmented asphalt over the entire bikeway surface can be used effectively and attractively as a substitute for painted lines, though the expense is often prohibitive. The roadway surface should not be painted with street paint because it is extremely slippery when wet.

FIGURE 59

Reflectorized Pavement Markers ("Dots")

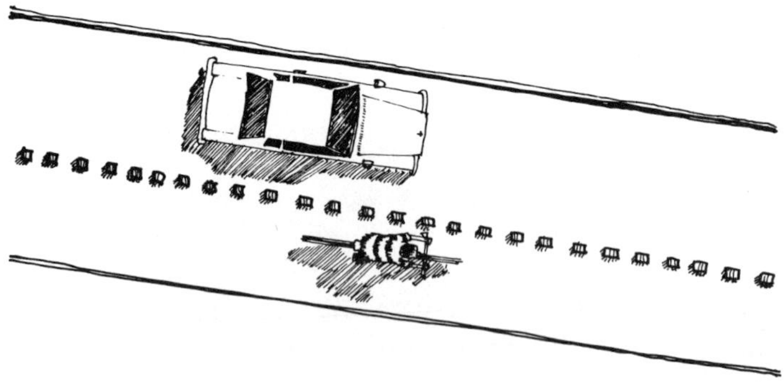

Reflectorized pavement markers ("dots") are most often used to reinforce painted line barriers; they provide reflected light at night, and automobile drivers hear a thumping sound when they cross them.

Bikeway Signs

FIGURE 60

Bike Route Sign

This sign is used to guide bicyclists along a designated bikeway and is accepted by the U.S. Department of Transportation, Federal Highway Administration. (*Manual on Uniform Traffic Control Devices*)

FIGURE 61

Bike Crossing Sign

This sign warns traffic in advance of a bikeway crossing point and is accepted by the U.S. Department of Transportation, Federal Highway Administration. (*Manual on Uniform Traffic Control Devices*)

Drainage Grate

FIGURE 62

Drainage Grate

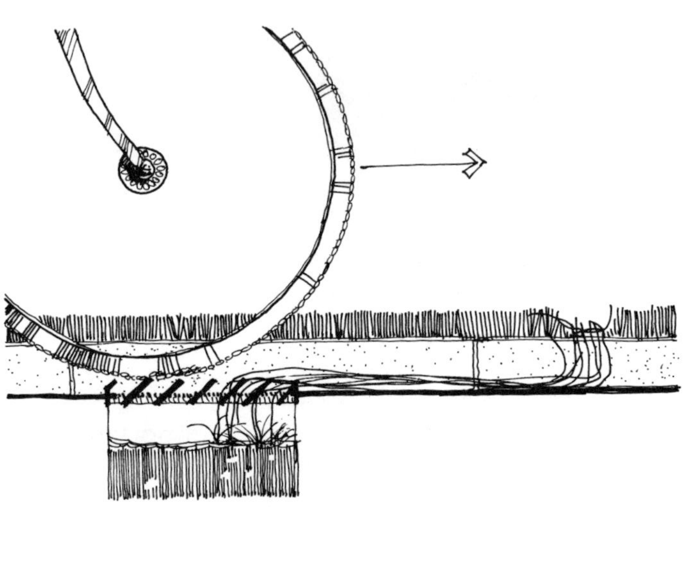

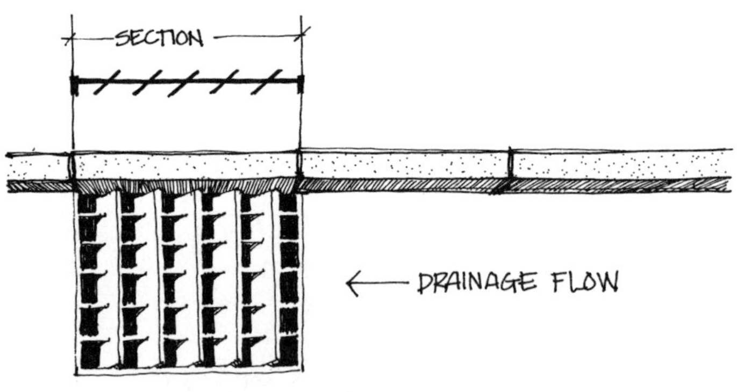

This new drainage grate used in Oregon is hydraulically superior to "parallel bar" grates and is also safe for bicyclists to ride over. (Federal Highway Administration Regional Hydraulic Engineer, Portland, Oregon)

Bikeway Paving

FIGURE 63

Asphalt Bikeway Surface

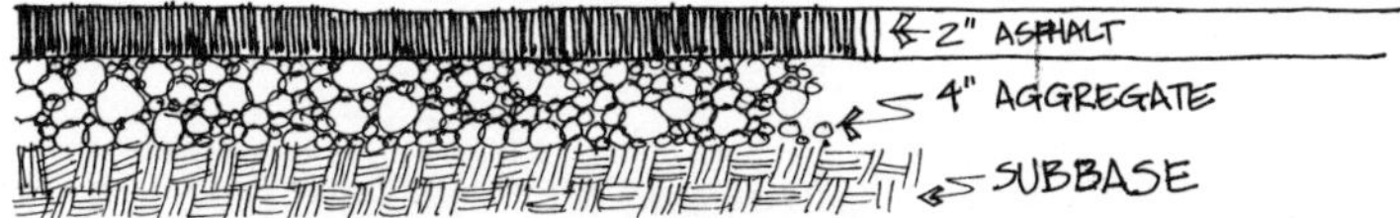

Asphalt is the most common Class I bikeway surface, as it is relatively inexpensive in construction and maintenance and can support maintenance vehicles. (Walter L. Cook, *Bike Trails and Facilities*)

FIGURE 64

Concrete Bikeway Surface

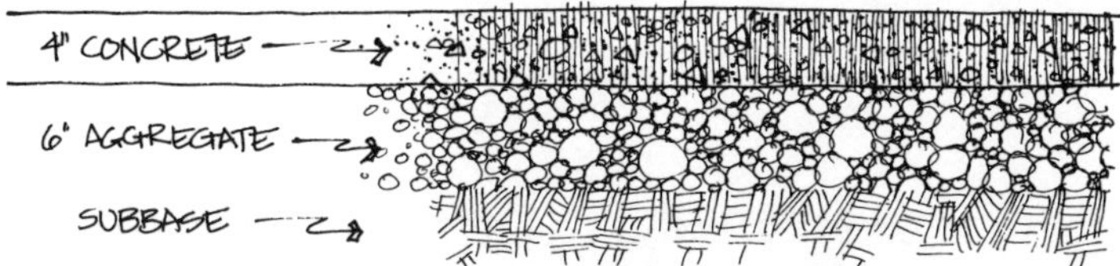

Concrete is recommended for high-use urban bikeways which are located above the curb and adjacent to concrete pedestrian walks. It will stand up to the high-use conditions better than asphalt. (Walter L. Cook, *Bike Trails and Facilities*)

FIGURE 65

Stabilized Earth Bikeway Surface

Stabilized earth is one of the least expensive surfaces and could prove valuable for hiking and biking trails, especially in dry climates.

Bicycle Parking Facilities

FIGURE 66

Bicycle Locker at BART Station

Bicycle lockers are available for 25¢ per day or $5 per month at some BART stations, California.

FIGURE 67

Steel Post with Chain at BART Station

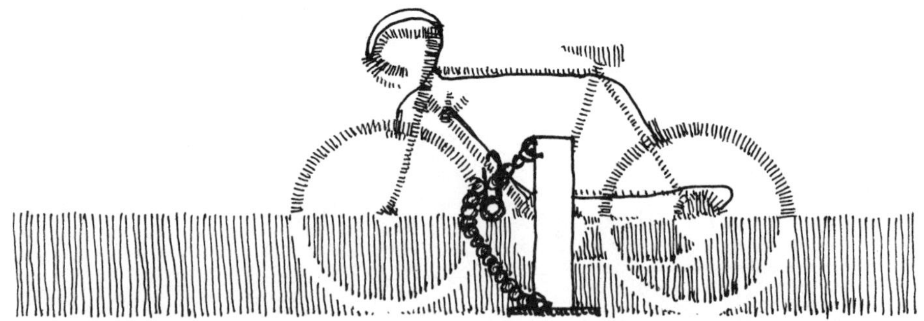

Steel post with chain at BART stations, California. The user provides the lock for this simple bicycle rack with case-hardened chain.

FIGURE 68

Coin-operated Bicycle Rack

Coin-operated bicycle rack at Golden Gate Bridge District, San Francisco, California. The rack locks the frame and both wheels without the need for a separate lock.

FIGURE 69

Hooks for vertical bicycle storage

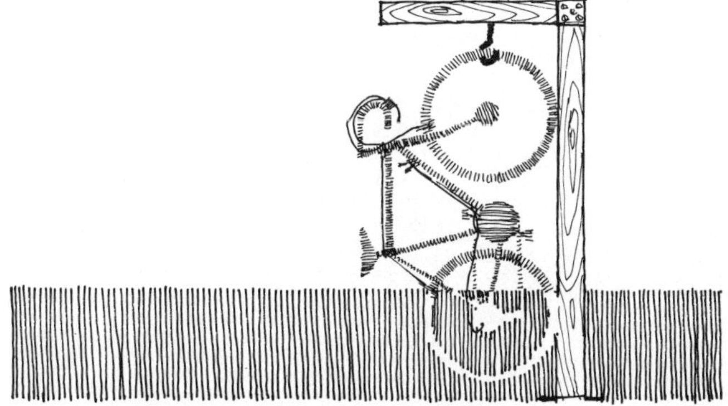

Hooks for vertical bicycle storage: this is one of the simplest racks, though it provides no device onto which the bicycle can be locked. An eyebolt could be installed on the vertical wall near the bicycle's crank and a cable could be fastened there. (*Voorzieningen voor het parkeren en onderbregen van fietsen,* Nederland)

It is important to remember that a survey of bikeway standardization should by no means be considered an absolute guideline for establishing engineering criteria for bicycles at the community or local level. Each community should first consider local factors such as site characteristics, existing traffic flow, and bicycle usage in the area before adopting policies for bikeway design. This discussion of bikeway standardization should provide an orientation for the engineer/designer, so that further study can be carried on at the local level.

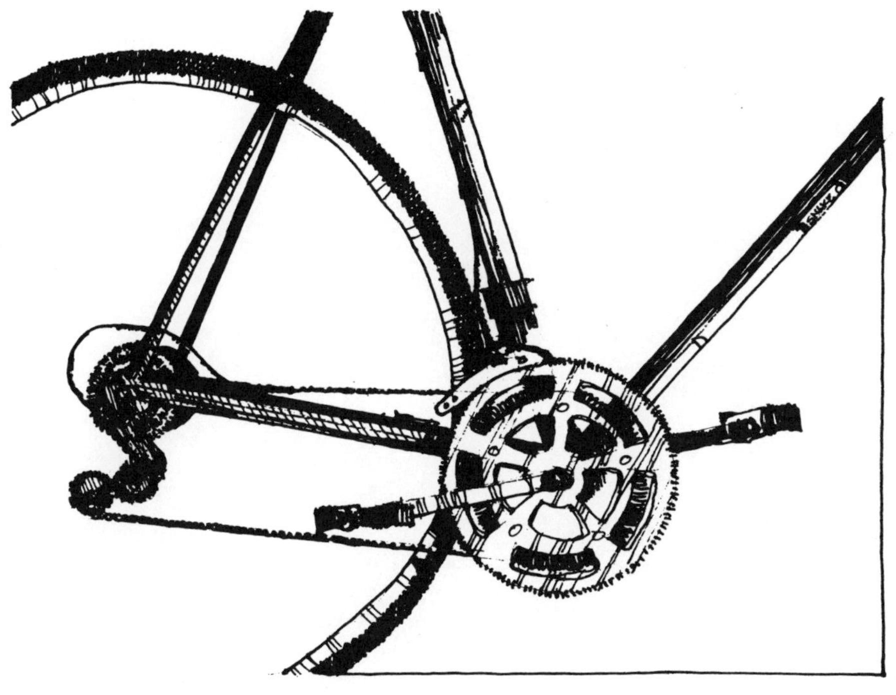

BICYCLE BILLS

During the recent energy crisis, Peter M. Flanigan, an assistant to the president, made the comment, "The United States is not going back to the cold, the dark, and the bicycle."[62] Such remarks are indicative of why bicyclists face inadequate saftey programs, ambiguous state and local ordinances, and insufficient funding or facilities.

Until very recently bicyclists in this country have been for the most part citizens between the ages of five and twenty, largely without voice or vote. It is for this reason that legislative revamping of outdated state and federal bicycle laws has been almost nonexistent and bicycle funding allocations for state and local governments very limited. Hopefully, as the adult population of bicycle riders grows, legislation which is more favorable to the needs of these people will be enacted. One indication that interest in bicycling is growing is the amount of recent bicycle legislation around the country, including more than 250 bills introduced in forty-three states during 1973 alone.[63]

The following cross-section of recent legislation reveals some of the directions being pursued in response to bicyclists' increased lobbying efforts.

H.B. 2282 (2/13/73) Arizona. Provides that the state highway director design and construct a system of bicycle pathways and foot pathways adjacent to certain state highways in cooperation with the federal aid highway program.

S. 115 (1/10/73) Connecticut. Provides for enactment of a law establishing bicycle lanes on state highways together with appropriate road signs to provide for the safety of bicyclists and motorists.

H.B. 3095 (1/3/73) Massachusetts. Provides for setting up a joint board of three state departments to develop a master plan and a method of financing a system of highway-related trails, including bicycle paths and hiking trails.

S. 198 (2/27/73) Michigan. Directs the State Highway Department to construct a system of intra-city bicycle paths between Detroit and Sault Sainte Marie, paralleling state or federal highways, and to assure their maintenance and repair.

H.B. L196 (1/16/73) Nebraska. Authorizes use of Highway Allocation Funds by cities and counties for the establishment of bicycle trails and footpaths wherever a highway, road, or street is being built, rebuilt, or relocated.[64]

As all bicycle facility construction is ultimately dependent upon both state and federal money, legislation concerning the collection and dispersal of state and federal revenues is of major importance. In 1971, the State of Oregon passed House Bill 1700, which calls for 1 percent of all funds received by the State Highway Commission to be spent on the establishment and construction of footpaths and bicycle trails. H.B. 1700 is not intended to encompass total funding for bikeway construction, but it does serve as an example of a model bill that is currently aiding bikeway construction.[65] Other states, such as California, are experimenting with similar types of funding bills which allow a specified percent of their state gasoline taxes to be spent on bikeway construction.[66]

It should be remembered that communities which seriously intend to establish comprehensive bikeways must not rely totally on state funds to build them. Matching funds must be made available for bikeway construction either in federal grants which can be received through the Department of the Interior or from general city funds.

In order to obtain money for bikeway construction, the city or community must first formulate a city-wide comprehensive plan, as suggested in the planning model. Only after the city council has ratified plans which have been produced by the neighborhood cell task forces can the process of funding and implementation begin at the city level. Priorities must be quickly established with regard to where city tax revenues will be spent in the future.

Clarification of local and state ordinances regarding bicyclists would facilitate speedier allocation of revenues and strengthen ineffectual safety information programs. Many attempts to educate the public concerning bicycle safety rules and city regulations have failed because their campaigns were designed for the entire population, rather than for the individual (through vehicles such as the Cycle Pack).

BICYCLE SAFETY PROGRAMS

Just as responsive planning and design must begin within neighborhood cell groups. attempts to inform the public of existing safety rules and city regulations are most effectively instituted at the cell level. In Santa Clara County, California, bicycle safety instruction is begun in the elementary schools and carried on to the high schools through the use of a series of safety manuals. These manuals are made available free of charge to all schools in the community, both public and private.

The manuals include a teacher's lesson manual called *Safety Education* and a student textbook titled *Bicycle Rules of the Road.* The manuals are designed so that a teacher can develop a lesson plan around bicycle safety, for example, incorporating bicycle history with studies in American history. The scope of the manuals expands as the type and extent of student bicycle use grows in diversity and range. The junior high school text, titled *An Operator's Guide to Safe and Enjoyable Bicycling* is included in driver education classes, giving the student a wider perspective of the complementary and interacting roles played by each transit mode within a comprehensive system.

Because it is difficult to reach the large section of the adult population not connected with the schools, public service announcements on television and radio and in printed media have been utilized in the effort to teach bicycle safety. However, adult bicyclists might best be reached through channels within the neighborhood bicycle cell organization. Since it is hoped that the process of cell design will be a cooperative venture within the family and neighborhood groups, bicycle safety and legal education could also begin within the home, with children bringing home safety literature from school. In more industrial or commercial neighborhood cells, information on safety problems and specific ordinances concerning the cell could be disseminated through the neighborhood cell's task force.

BICYCLE RULES AND REGULATIONS

A great deal of the difficulty in formulating bicycle rules and regulations arises from confusion about the bicycle's relationship to automobiles and pedestrians. The bicyclist is simply not a motorist or a pedestrian and consequently requires rules and regulations different from those of other transit forms. Navigation codes regulate different vessels using the waterways according to differences in their scale, speed, and flexibility, and a similar method could be adapted to differentiate forms of land transit.

One means of instituting more egalitarian treatment of all land-based transit modes would be the requirement of a bicycle operator's license. The licensing procedure would familiarize both professionals, such as local police officers, and citizens with their obligations. At present, the law does not require the bicyclist to carry a driver's license while riding. It is a mistake on the part of the officer to ask the bicyclist for a license; any resulting ticket may count against the bicyclist's automobile driving record and could conceivably raise his automobile insurance rates.[67] If the situation arises, some substitute form of identification should be respectfully

FIGURE 70

Importance of Traffic Safety Will Grow

Traffic safety will become even more important as the number of bicycle enthusiasts grows.

offered in its place. A bicycle operator's license would be a good alternative.

In Pendleton, Oregon, a local ordinance states that all bicycle operators under the age of eighteen years shall take a written examination on local bicycle safety laws and pass with a grade of 80 percent.[68] This ordinance is a significant step in the direction of licensing competent bicycle operators in the state. As bicycles assume an important role within future transportation systems, it will be imperative that a high degree of safety and alertness is insured. It seems that uniform licensing of bicycle operators of all ages will be inevitable. Separate licenses are already required in several states to operate motorcycles, independent of the motor-cyclist's previous automobile licensing. California, Oregon, and Idaho all have such laws. In California, exploration of a state-wide bicycle licensing program is under way. The proposed program would make it mandatory for all bicycles within the state to be licensed with the California Department of Motor Vehicles.

The revenues generated by licensing programs would make it possible to allocate funds to city and county agencies for improving bicycle registration and safety programs in their jurisdictions. The inclusion of bicycle licenses on all bicycles would greatly reduce bicycle theft. Many theft-prevention measures have been tried by local communities, but because licensing is local, interception of stolen bicycles after they leave the community is impossible. The introduction of state bicycling registration programs as proposed by the State of California could help reduce the number of stolen bicycles, increase the number retrieved, and insure greater security and probability of use.

LEGAL JUSTIFICATION OF BIKEWAYS

Of growing concern to bicyclists is the loss of their right to use public roadways. Bicyclists fear that when municipalities begin to provide separate bikeways, bicycles will be prohibited from using public streets or will be limited only to certain specified streets. Some cities already have ordinances prohibiting bicycling in their central business district.

> The worst thing cycling is promoting is the bikeway movement. . . . Cycling in the U.S. and Canada is an anomaly, occupying a far different and inferior position than it does anywhere else in the world, which should suggest caution and additional study before every promo-tional attempt.[69]

Some bicyclists contend that bicyclists' freedom to travel on the public highway is being undermined and they vigorously oppose any specific law that would limit cycling.

Although the exclusion of bicyclists from the public roadways is not legal, it is being done. The denial of public street access to bicyclists is of primary interest to all those who support bicycle transit systems. The League of American Wheelmen opposes any bikeway development plan which would deny bicyclists the use of public streets, roads, and highway facilities. The League only supports bike paths as separate facilities where no public roads exist, on bridges to bypass or parallel limited-access highways, or in special recreation and park areas.[70] This sentiment is also expressed in *Bike World Magazine:*

> Some highways have been specially constructed for high-speed traffic, with limited access and user rights, but this limitation does not conflict with the principle of free use of the highway system as long as the pre-existing slow-speed, unlimited access roads serve other traffic. Only in those few areas where a high-speed freeway has wiped out the existing road does infringement exist, and these cases have been solved in principle (though not publicly acknowledged) by either permitting access or constructing a frontage road or path.[71]

Another legal question concerning bikeways is the exclusive or dominant use of roadways by one form of transportation. The City of Milwaukee, Wisconsin, designated special traffic lanes for bus use only, and the State Supreme Court later ruled that exclusive bus lanes are illegal. The court ruled that

> although the idea of exclusive lanes for certain vehicles might be in the public interest, municipalities have no right on their own to establish a function which the legislature has not seen fit to delegate . . . and that until the legislature allowed municipalities to set aside exclusive lanes for certain vehicles, the concept was illegal.[72]

FIGURE 71

Equestrian and Bicycle Ordinance, 1886

The introduction of the bicycle in Oregon "so aggravated the perils of traffic that the Oregon legislature passed a law to the effect that cyclists should halt whenever they approached within a hundred yards of a team, dismount, and remain standing until the horses had passed. In commenting on this law, the *Daily Astorian,* July 13, 1886, declared 'This law may be a good one, but it does not go far enough; it should be amended so as to compel the bicyclist to take off his hat and remain uncovered while the driver of the team is passing.'" (*Oregon History Quarterly,* vol. 61)

This ruling would also apply to lanes set aside for the exclusive use of bicycles in Wisconsin. However, laws in other states may allow municipalities to develop this type of system.

The exclusive-use situation can be viewed from another perspective. If all or part of a roadway cannot be set aside for the exclusive use of one type vehicle in certain states, arguments can be made that most existing roads have been set up for the exclusive use of the automobile. The speed limits, street designs, and rules of the road often act effectively to exclude vehicles that are not high-powered and thereby violate the spirit of the law.

An important aspect of the exclusive-use concept is the idea of establishing times when different types of vehicles can use a roadway or a part of it. Commuter buses on some of the more congested highways have been allotted certain lanes for their exclusive use during rush hours; Golden Gate Park in San Francisco is successfully closed to automobile use on Sundays, but not to bicycles. Temporal regulations are not considered restrictive in an exclusive sense, as long as reasonable access at reasonable times is provided for all. The City of Davis, California, attempted to utilize a temporal regulation scheme on a narrow but busy stretch of roadway, with signs designating the hours during which parking was allowed. During busy commuting hours, no parking was allowed on the street. However, cars were often left illegally parked during the restricted periods, and bicyclists using the road at rush hours were forced to maneuver through hazardous conditions. Davis' solution was to prohibit parking on the street completely.[73]

Bicyclists are a minority group on public roads and thus must make a special effort to insure that their rights are not usurped and to re-establish them where they have been. To quote an official from the Department of the Interior, "It is a wonder there are any bike riders in this country at all, because so little has been done for them and so much against them."[74]

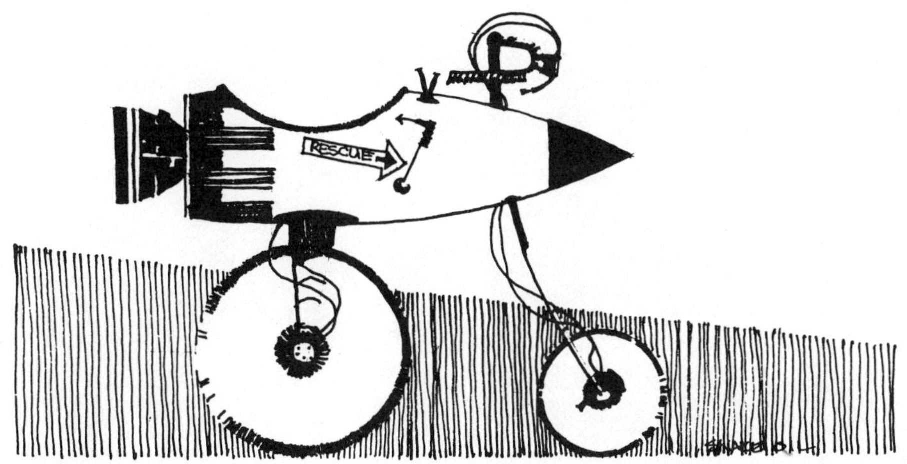

It has been nearly a century since the advent of the League of American Wheelmen and the bicycle clubs. Were it not for these groups and the pressures they brought to bear in quest of better, federally funded highways, modern transit would not have reached its present level of complexity. Bicycles would still not be considered serious forms of transit, and our national network of highways would be far less developed than it is.

The bicycle has suffered many setbacks in its progression into the modern transportation picture. Part of the difficulty has stemmed from the fact that many continue to view the bicycle exclusively as a recreational vehicle, to be used only in parkways or in nonurban open spaces. Regardless of the categorical dismissal of bicyclists and bicycles by legislatures, city governments, and planners of the past, the bicycle has been able to maintain itself as a valid means of transportation and must be considered so in any future transit planning.

It is difficult in the formulation of a book such as this to determine the prospective roles of highly specialized and rapidly changing modes of transportation and the extent to which they will be needed in the future. However, planning agencies have begun re-evaluating and exploring the potential of the bicycle as a transit

FIGURE 72

Some Things Won't Change Concerning Bicycles

Although the future of the bicycle is bright, certain aspects of bicycle transit will remain with us.

vehicle. This book has intended to identify those considerations most pertinent to the organization of comprehensive systems and the interaction of these systems with people and other modes of transit. The notion of comprehensive systems has been stressed in an attempt to clarify the process of investigation which must be thoroughly undertaken before any bikeway can become a truly integrated part of the transit and social environment.

The planning and design of bikeways is by nature very situational. Questions such as land-use planning, economics, the manner in which bikeways influence people's lives, and aesthetic and environmental considerations must all be taken into account before, during, and after the implementation of bikeways. If problem-solving strategies for each particular bikeway and transit situation are formulated and tested, planning will come to be viewed as a process rather than an exercise in placing and creating objects on the landscape. Through carefully observing and documenting the operational performance of new bikeways and transit systems, planners can evolve criteria for the development of future efforts. Collections of this kind of information can then be made available for further investigation, in the same way that this book has presented collected data.

This book can be used as a resource manual in a number of ways. Certain sections may provide a means of acquiring necessary information, while others may be valuable later on, in the context of a specific community's desired or available organization. For example, the Cycle Pack could prove an invaluable method of gaining preliminary information, while planning and design sections may be used at a later time or modified to suit situational needs. It should be remembered that the book was created as a complete manual, with each of its parts an extension of the others. While any part of it may be used alone or in conjunction with other bicycle or transportation information, its greatest benefits may be derived from using it as a comprehensive resource.

Further research should be carried out in a number of areas of investigation mentioned here. One area is the question of bikeway necessity within a given locale. Research could include organization of sales and licensing data; Cycle Pack surveys; traffic monitoring; and tapping public and community knowledge. Another topic is the interaction of bicycles and existing or future modes of transportation, including investigations into bicycle parking, modification of existing mass transit vehicles, and the rental of bicycles for one-way use. Research might also focus upon accommodations for long-range bicycling (regional and inter-regional), with study of hostels and touring and safety information concerning these areas. Pollution and the use of bicycles in polluted atmospheres is another possible area of concern. Research could include investigation of

FIGURE 73

Experimentation in Bicycle Design

In the future, we can expect experimentation in new bicycle designs.

the effects of plants in reducing pollution and the physical harm caused by bicycling in rush traffic. Other research might encompass the engineering of low-cost, elevated or subterranean bikeways, including design of bike racks and bike storage lockers, low-cost bikeway surfaces, and the use of asphalt. Current and future avenues for financing bikeways and other forms of transit, such as mixed-mode and mass transit, could be studied. This should include an in-depth examination of legislative process and control. Finally, research could cover the licensing of bicycles and the development of comprehensive safety instruction, with studies of methods for distributing bicycle safety information effectively.

Bicycling is an individual and personal form of transportation. Because it is mainly a localized activity, the elements most essential to bicycle planning lie within the neighborhood planning cell. Hopefully, concerned individuals within the cell whose daily activities make them most intimately aware of the necessity for bikeways will be given the opportunity to work with the professionals whose expertise and training can aid in producing responsive and aesthetically pleasing bicycle transit.

1. Liege, Belgium, Musée de la Vie Wallonne, *Cycles et Motocycles* [Cycles and motorcycles], p. 9.

2. Robert A. Smith, *A Social History of the Bicycle*, p. 1.

3. Musée de la Vie Wallonne, *Cycles et Motocycles*, p. 9.

4. Ibid., p. 11.

5. Smith, *Social History*, p. 4.

6. Ibid., p. 4.

7. Musée de la Vie Wallonne, *Cycles et Motocycles*, p. 15.

8. Smith, *Social History*, p. 5.

9. Musée de la Vie Wallonne, *Cycles et Motocycles*, p. 15.

10. Smith, *Social History*, p. 7.

11. Ibid., p. 8.

12. Ibid., p. 9.

13. Christopher Tunnard and Henry Hope Reed, *American Skyline*, p. 120.

14. Philip P. Mason, *The League of American Wheelmen and the Good-Roads Movement 1880-1905*, p. 57.

15. Hugh Myron Hoyt, Jr., *The Good Road Movement in Oregon 1900-1920*, p. 153.

16. Mason, *American Wheelmen*, p. 110.

17. Smith, *Social History*, p. 100.

18. Ibid., p. 99.

19. Ibid., p. 40.

20. Mason, *American Wheelmen*, p. 60.

21. Smith, *Social History*, pp. 242-43.

22. George a'Green, *This Great Club of Ours*, p. 67.

23. Smith, *Social History*, p. 247.

24. a'Green, *Great Club*, p. 67.

25. Barton-Aschman Associates, Inc., *The Bicycle*, p. 2.

26. Kenneth D. Cross and Richard de Mille, *Human Factors in Bicycle-Motor Vehicle Accidents*, p. 1.

27. Ibid.

28. David M. Eggleston, "Toward a Dual-Mode Bicycle Transportation System," p. 187.

29. Robert Sommer and Dale F. Lott, "Behavioral Evaluation of a Bikeway System," p. 2.

30. University of Wisconsin Department of Landscape Architecture, Environmental Awareness Center, *Milwaukee Stadium Freeway*, p. 129.

31. State of Indiana Department of Natural Resources and Indiana Central Bicycling Association, *Abandoned Railroad Rights-of-Way as Potential Bicycling and Hiking Trails*, pp. 3-4.

32. Inouye, Singer & Hodges, *Cross Marin Trail and Bicycle Route*, pp. 4-5.

33. Arizona Highway Department and Bivens & Associates, Inc., *Arizona Bikeways*, p. IV-69.

34. Wichita-Sedgwick County Metropolitan Area Planning Department, *The Canal Route Open Space Corridor*, p. 4.

35. Ibid., p. 31.

36. Charles Ramsdell, *Special Supplement to the HemisFair Edition of San Antonio*, p. 79.

37. Ibid., p. s13.

38. Katharine Campbell, William VanDyke, and Joan Williams, *Asphalt Empire*, p. iv.

39. Ibid., p. 73.

40. M. Lynn Stevenson, Presentation at Clean Air Act Public Hearing.

41. Institute of Transportation and Traffic Engineering, School of Engineering and Applied Science, University of California, Los Angeles, *Bikeway Planning Criteria and Guidelines*, p. 151.

42. Barton-Aschman Associates, Inc., *The Bicycle*, pp. 21-23.

43. Hart, Krivatsy, Stubee, *BART/Trails*, p. 22.

44. Bay Area Rapid Transit District, "Bikes on BART Test Program," pp. 2-3.

45. Koninklijke Nederlandsche Toeristenbond ANWB, *Fietspaden en -oversteekplaatsen* [Bicycle paths and cycle-crossings], p. 60.

46. Gary O. Robinette, *Plants/People/and Environmental Quality*, p. 87.

47. Ibid., p. 86.

48. Rémy Chauvin, *The World of an Insect*, p. 32.

49. Robinette, *Plants/People,* p. 75.

50. Ibid., p. 78.

51. Robert B. Deering and Frederick A. Brooks, "The Effect of Plant Material Upon the Microclimate of House and Garden," p. 165.

52. Robinette, *Plants/People,* p. 97.

53. Ibrahim Joseph Hindawi, *Air Pollution Injury to Vegetation,* p. 3.

54. Michael D. Everett, "Roadside Air Pollution Hazards in Recreational Land Use Planning," p. 83.

55. Ibid., p. 84.

56. Ibid., p. 86.

57. Robinette, *Plants/People,* p. 54.

58. Malcolm C. Shurleff, "How Air Pollution Affects Tree Varieties," p. 20.

59. Lloyd N. Popish and Roger B. Lytel, *A Study of Bicycle-Motor Vehicle Accidents,* p. 42.

60. Gruppe Radwegebau, *Radverkehrsanlagen Richtlinien* [Guidelines for bicycle traffic layouts].

61. Institute of Transportation, *Bikeway Criteria,* p. 27.

62. Friends for Bikecology, "An Anti-Bikeway Movement."

63. Bicycle Institute of America, "State Bicycle Legislation Proliferating."

64. Bicycle Institute of America, *Summary of 1973 Bikeway Legislation.*

65. Oregon State House Bill 1700, p. 338.

66. California State Senate Bill No. 821, pp. 1517-20.

67. Bob Boethling, ed., *The Bicycle Book,* p. 30.

68. City of Pendleton, Oregon, "Pendleton Ordinances," S. 55.

69. John Forester, "What About Bikeways?" p. 36.

70. Clifford L. Franz and A. Fred DeLong, "California Traffic Safety To Include All Road Users," pp. 7-8.

71. Forester, "What About Bikeways?" p. 37.

72. City of Milwaukee Bureau of Traffic Engineering and Electrical Service, D.P.W., *A Proposed System of Bicycle Commuting Routes and Comments on the Use of Bicycles on Sidewalks,* p. 1.

73. Sommer and Lott, "Behavioral Evaluation," p. 7.
74. Ralph Hanneman, "Success Story: The Growth of
'Bikeways.' "

BIKEWAY STUDIES

In the course of our research, we found some bicycle studies more valuable than others. Throughout the list, those sources which provide highly pertinent information are indicated with an asterisk (*). Mailing addresses have been included in most cases.

Arizona

*Arizona Highway Department and Bivens & Associates. 1973. *Arizona Bikeways: A Comprehensive Bicycle Program for Arizona.* Phoenix, Arizona. (Arizona Highway Department, 1739 W. Jackson St., Phoenix, AZ 85007)
__________. 1973. *Arizona Bikeways Summary: A Comprehensive Bicycle Program for Arizona.* Phoenix, Arizona. (Arizona Highway Department, 1739 W. Jackson St., Phoenix, AZ 85007)
Arizona State Highway Commission. 1967. *Arizona Bicyclist's Manual.* 2d ed., Phoenix, Arizona. (Traffic Safety Division, Arizona Highway Department, 2339 No. 20th Ave., Phoenix, AZ 85009)
Tempe Planning Department. 1972. *Tempe Bikeway Study: Background.* Tempe, Arizona. (Tempe Planning Department, P.O. Box 5002, Tempe, AZ 85281)
__________. 1973. *Tempe Bikeway Study: Preliminary Plans and Recommendations.* Tempe, Arizona. (Tempe Planning Department, P.O. Box 5002, Tempe, AZ 85281)

California

*A C Transit. "Bus Bridges the Bay for Bicycle Riders." *Transit Times* 14(1):10. (A C Transit, 508 16th St., Oakland, CA 93512)

*Bay Area Rapid Transit District. 1970. "Bicycle Policies and Facilities at BART." Oakland, California. (Bay Area Rapid Transit District, 800 Madison St., Oakland, CA 94607)

————. 1974. "Bikes on Bart Test Program." Oakland, California. (Bay Area Rapid Transit District, 800 Madison St., Oakland, CA 94607)

*Berkeley City Planning Department. 1971. *Bicycles in Berkeley: A Background Report.* Berkeley, California. (Berkeley Public Works Department, City Hall, Berkeley, CA 94704)

*Bicycle Safety for Santa Clara County Project. 1973. *Safety Education: Grade Five.* San Jose, California. (Bicycle Safety for Santa Clara County Project, Rm. 103, 460 Park Ave., San Jose, CA 95110)

———— and Diridon Research Corp. 1973. *Public Attitude Survey for the Bicycle Safety for Santa Clara County Project Survey Data Analysis.* San Jose, California. (Bicycle Safety for Santa Clara County Project, Rm. 103, 460 Park Ave., San Jose, CA 95110)

*Boethling, Bob, ed. 1971. *The Bicycle Book.* Los Angeles: UCLA Alumni Association. (UCLA Sierra Club, UCLA Box 118, 308 Westwood Plaza, Los Angeles, CA 90024)

*California Department of Transportation. 1973. "Summary of Bikeway Funding in California." Sacramento, California.

————. 1974. *Highway Project Development: Highway Design Bike Routes.* Sacramento, California. (California Department of Transportation, Division of Highways, 1120 N St., Sacramento, CA 95814)

*California State Senate. 1973-1974 Regular Session. "Public Utilities: Transportation Development." ch. 740, Senate Bill No. 821, Sacramento, California.

*————. 1972. "State Highways: Bicycle Facilities." ch. 1092, Senate Bill No. 36, Sacramento, California.

City of Concord Planning Department. 1972. *City of Concord Trails Plan.* Concord, California. (City of Concord Planning Department, Civic Center, 1950 Parkside, Concord, CA 94519)

City of Davis Department of Public Works. 1970. "New Bike Lanes in Relationship to Change in Subdivision Ordinance and General Plan Standards." Davis, California. (City of Davis Planning Department, City Hall, Davis, CA 95616)

*City of Palo Alto. 1972. *A Study of Attitudes and Awareness of Bicycling Procedures and Safety: A Work Task for Developing Continuing Comprehensive Public Education and Information Program for Bicycling.* Palo Alto, California. (City of Palo Alto Department of Public Works, City Hall, Palo Alto, CA 94301)

————. 1973. "The Urban Bicycle Route System." Palo Alto, California. (City of Palo Alto Department of Public Works, City Hall, Palo Alto, CA 94301)

———— and Mortimer Associates. *Official Bicycle Manual.* Palo Alto, California. (City of Palo Alto Department of Public Works, City Hall, Palo Alto, CA 94301)

City and County of San Francisco, Department of Public Works, Bureau of Engineering. "San Francisco Bicycle Routes." San Francisco, California. (City and County of San Francisco, Department of Public Works, Division of Traffic, 460 McAllister St., San Francisco, CA 94102)

*City of San Jose Department of Public Works. 1973. *A Bicycle Safety Program for Parents' Organizations.* San Jose, California. (Bicycle Safety Project for Santa Clara County, Rm. 103, 460 Park Ave., San Jose, CA 95110)

*————. 1974. *An Operator's Guide to Safe and Enjoyable Bicycling.* San Jose, California. (Bicycle Safety Project for Santa Clara County, Rm. 103, 460 Park Ave., San Jose, CA 95110)

*Cross, Kenneth D., and de Mille, Richard. 1973. *Human Factors in Bicycle-Motor Vehicle Accidents.* Santa Barbara, California. (City of Santa Barbara Department of Public Works, P.O. Box 33, Santa Barbara, CA 93102)

Davis Police Department. City of Davis Bicycle Regulations. Davis, California. (City of Davis Planning Department, City Hall, Davis, CA 95616)

*De Leuw, Cather & Co. 1972. *City of Davis, University of California: Bicycle Circulation and Safety Study.* Davis, California. (City of Davis Planning Department, City Hall, Davis, CA 95616)

*Eggleston, David M. 1973. "Recent Progress in Dual-Mode Bicycle Transportation." San Diego, California: California State University.

*______. 1973. "Toward a Dual-Mode Bicycle Transportation System." In *Proceedings of the Pedestrian/Bicycle Planning and Design Seminar, San Francisco, 1972.* Berkeley, California: Metropolitan Association of Urban Designers and Environmental Planners (MAUDEP) and the Institute of Transportation and Traffic Engineering. (ASUC Store, University of California, Berkeley, CA 94720)

*Hart, Krivatsy, Stubee. 1974. *BART/Trails: A Study of the Commuter and Recreational Trail Potential to Bay Area Rapid Transit System.* Oakland, California: U.S. Department of Transportation, Office of Assistant Secretary for Environment, Safety and Consumer Affairs. (Hart, Krivatsy, Stubee, Planning Architecture Environmental Design, 220 Jackson St., San Francisco, CA 94111)

*Inouye, Singer & Hodges. 1973. *Cross Marin Trail and Bicycle Route: Preliminary Master Plan Report.* San Rafael, California: Department of Parks and Recreation, County of Marin and Marin Municipal Water District.

*Institute of Transportation and Traffic Engineering. 1972. *Bikeway Planning Criteria and Guidelines.* Los Angeles: School of Engineering and Applied Science, University of California.

Joske, Pierre. 1970. "Bike Path: Progress Report." San Rafael, California: Marin County Parks and Recreation Department. (County of Marin, Department of Parks and Recreation, Civic Center, San Rafael, CA 94903)

*Local Transit Study Committee. 1971. *Berkeley Bikeways Plan.* Berkeley, California. (Berkeley Public Works Department, City Hall, Berkeley, CA 94704)

*Marsin, Alice, and Silberstein, Jane. 1973. *City of Santa Barbara Bikeway Planning Surveys.* Santa Barbara, California: City of Santa Barbara Public Works Department, Traffic Division. (City of Santa Barbara Public Works Department, P.O. Box 33, Santa Barbara, CA 93102)

Orange County Planning Department. 1974. *Master Plan of County-wide Bikeways for Orange County Amendment No. 4.* Santa Ana, California. (Orange County Planning Commission, P.O. Box 4108, Santa Ana, CA 92701)

Popish, Lloyd N., and Lytel, Rober B. 1973. *A Study of Bicycle-Motor Vehicle Accidents.* City of Santa Barbara, California.

Proceedings of the Pedestrian/Bicycle Planning and Design Seminar, San Francisco, 1972. Berkeley, California: Metropolitan Association of Urban Designers and Environmental Planners (MAUDEP) and the Institute of Transportation and Traffic Engineering. (ASUC Store, University of California, Berkeley, CA 94720)

Regional Parkland and Trails Map. 1973. Overview, Oakland, California. (East Bay Regional Park District, 11500 Skyline Blvd., Oakland, CA 94619)

Regional Trails. 1973. Overview, Oakland, California. (East Bay Regional Park District, 11500 Skyline Blvd., Oakland, CA 94619)

San Francisco Department of City Planning. 1971. *The Urban Design Plan for the Comprehensive Plan of San Francisco.* San Francisco. (San Francisco Department of City Planning, 100 Larkin St., San Francisco, CA 94102)

__________. 1972. *Plan for Transportation: The Comprehensive Plan of San Francisco.* San Francisco. (San Francisco Department of City Planning, 100 Larkin St., San Francisco, CA 94102)

San Mateo County Bikeways Committee and San Mateo County Planning Commission. 1972. *Bicycle Planning in San Mateo County: Progress Report 1.* Redwood City, California. (San Mateo County Planning Commission, County Government Center, Redwood City, CA 94063)

Santa Barbara Public Works Department. 1974. *The Proposed Bikeway Master Plan.* Santa Barbara, California. (Santa Barbara Public Works Department, P.O. Box 33, Santa Barbara, CA 93102)

*Santa Clara County Bicycle Safety Project. 1973. *Bicycle Rules of the Road.* San Jose, California. (Bicycle Safety Project for Santa Clara County, Rm. 103, 460 Park Ave., San Jose, CA 95110)

Shanteau, Bob. *What To Do With the Extra Part When You're All Finished: Guide to Bike Repair and Maintenance.* San Jose, California: Bicycle Safety Project for Santa Clara County. (Bicycle Safety Project for Santa Clara County, Rm. 103, 460 Park Ave., San Jose, CA 95110)

*Sommer, Robert, and Lott, Dale F. 1971. "Bikeway in Action:
    The Davis Experience." Davis, California: Department of
    Psychology, University of California. (City of Davis Planning
    Department, City Hall, Davis, CA 95616)
————. 1972. "Behavioral Evaluation of a Bikeway System."
    Davis, California: Department of Psychology, University of
    California. (City of Davis Planning Department, City Hall,
    Davis, CA 95616)
State of California Resources Agency, Department of Water
    Resources. "California Aqueduct Bikeway: A State Water
    Project Benefit." Byron, California. (State of California,
    Department of Water Resources, Route 1, Box 39, Byron, CA
    94514)
Stevenson, M. Lynn. 1973. "Tour of Possible Bicycle Parkway-
    Arterials in Berkeley." California Association of Bicycling
    Organizations and Energy and Environment Program, October
    25.

Colorado

*Citizens' Advisory Committee on Trails. 1971. *Bikeways for Lake-
    wood.* Lakewood, Colorado. (The City of Lakewood, Depart-
    ment of Community Services, 1580 Yarrow St., Lakewood,
    CO 80215)
City of Boulder Transportation Division. 1972. "The Boulder
    Valley Bikeway Plan Proposed." Boulder, Colorado. (City of
    Boulder Transportation Division, P.O. Box 791, Boulder, CO
    80302)
*Denver Planning Office. 1973. *The Bike Plan.* Denver, Colorado.
    (City of Denver Planning Office, Rm. 300, 1445 Cleveland Pl.,
    Denver, CO 80202)

Florida

*Jacksonville Area Planning Board. 1973. *Bike System Plan.*
    Jacksonville, Florida. (City of Jacksonville Department of
    Public Works, Engineering Division, City Hall, Jacksonville,
    FL 33202)

*Metropolitan Dade County Planning Department. 1972. *Proposed Dade Bikeway Plan.* Miami, Florida. (Metropolitan Dade County Planning Department, 702 Justice Building, Miami, FL 33125)

*South Florida Regional Planning Council and Florida Department of Natural Resources. 1972. *Bicycle Safety Trails in South Florida: Proceedings from a Conference.* Coral Gables, Florida. (South Florida Regional Planning Council, P.O. Box 8083, Coral Gables, FL 33124)

State of Florida Department of Natural Resources and Department of Transportation. 1974. *Guidelines for Implementing Florida's Bicycle Trail Program.* Tallahassee, Florida. (State of Florida Department of Natural Resources, Larson Building, Tallahassee, FL 32304)

University of Florida Landscape Architecture Department. *Guidelines for a Bikeway System.* Gainesville, Florida. (University of Florida, Division of Planning and Analysis, Gainesville, FL 32601)

Georgia

*Barton-Aschman Associates. 1973. *The Bicycle: A Plan and Program for Its Use as a Mode of Transportation and Recreation.* Atlanta, Georgia: Atlanta Metropolitan Region, Atlanta Regional Commission. (Barton-Aschman Associates, Inc. 1821 University Ave., St. Paul, MN 55104)

_______. 1973. *The Bicycle: A Plan and Program for Its Use as a Mode of Transportation and Recreation, Technical Appendix.* Atlanta, Georgia: Atlanta Metropolitan Region, Atlanta Regional Commission. (Barton-Aschman Associates, Inc., 1821 University Ave., St. Paul, MN 55104)

Idaho

BMTS Subcommittee for Bicycle, Pedestrian, and Equestrian Paths.
  1973. "Goals and Objectives for Bicycle Path Planning in the
  Boise Metropolitan Area." Boise, Idaho. (ADA Council of
  Governments, 525 W. Jefferson, Boise, ID 83702)

Illinois

*Baerwald, John E. 1973. *Bicycle and Campus Transportation.*
  Urbana, Illinois: Highway Traffic Safety Center, University
  of Illinois. (Highway Traffic Safety Center, University of
  Illinois, 418 Engineering Hall, Urbana, IL 61801)
_______. 1973. "Operation and Maintenance of Campus Bicycle
  Facilities." Urbana, Illinois: Highway Traffic Safety Center,
  University of Illinois. (Highway Traffic Safety Center, Univer-
  sity of Illinois, 418 Engineering Hall, Urbana, IL 61801)
Chicago Park District. 1971. *Cyclist Guide: Chicago Bicycle Route
  System.* Chicago. (City of Chicago Department of Develop-
  ment and Planning, Rm. 1000, City Hall, Chicago, IL 60602)
*City of Chicago Department of Development and Planning. 1971.
  Guidelines for a Comprehensive Bicycle Route System.
  Chicago. (City of Chicago Department of Streets and Sanita-
  tion, Bureau of Street Traffic, Rm. 703, City Hall, Chicago,
  IL 60602)
La Plante, John N. 1973. *Chicago Bikeway Progress Report.*
  Chicago. (City of Chicago Department of Development and
  Planning, Rm. 1000, City Hall, Chicago, IL 60602)
*Lewis, John W. 1971. *Illinois Bicycle Rules of the Road.* Chicago.
  (City of Chicago Department of Development and Planning,
  Rm. 1000, City Hall, Chicago, IL 60602)

Indiana

*State of Indiana Department of Natural Resources and Indiana
Central Bicycling Association, 1972. *Abandoned Railroad
Rights-of Way as Potential Bicycling and Hiking Trails.* Indi-
anapolis, Indiana. (State of Indiana Department of Natural
Resources, 612 State Office Building, Indianapolis, IN 46204)

Iowa

*Committee of Citizens for a Better Iowa City. 1969. *Hawkeye
Area Bikeway System: A Proposal From the Bikeways Subcommittee of Project Green.* Iowa City, Iowa. (City of Iowa
City, Civic Center, 410 E. Washington St., Iowa City, IA
52240)
Des Plaines Police Department. 1972. "Proposal for Organization
and Function of Bicycle Safety Program." Des Plaines, Iowa.
(City Plan and Zoning Commission, Argonne Armory, E. 1st
and Des Moines Sts., Des Moines, IA 50309)

Kansas

*Wichita-Sedgwick County Metropolitan Area Planning Depart-
ment. 1970. *The Canal Route Open Space Corridor: A Joint
Development Plan for I-35W Through Wichita, Kansas.*
Wichita, Kansas. (Wichita-Sedgwick County Metropolitan Area
Planning Department, City Building Annex, 104 So. Main St.,
Wichita, KS 67202)

Kentucky

City-County Planning. 1972. "Evaluation of Bicycle Facilities and
   Needs." Lexington, Kentucky. (Lexington-Fayette County
   Planning Commission, 227 No. Upper St., Lexington, KY
   40507)

Massachusetts

Boston Redevelopment Authority. 1973. *Boston, Brookline,
   Cambridge Metropolitan Bike Study: Pilot Route Plan.*
   Boston, Massachusetts. (Boston Redevelopment Authority,
   City Hall, Rm. 900, 1 City Hall Square, Boston, MA 02201)

Michigan

City of Detroit City Plan Commission. *Detroit 1990/An Urban
   Design Concept for the Inner City.* Detroit Michigan. (City of
   Detroit, City Plan Commission, 8th Floor City-County
   Building, 2 Woodward Ave., Detroit, MI 48226)
*Smith, Haldon L. 1973. *City of Ann Arbor Bike Path Study.* Ann
   Arbor, Michigan: Department of Housing and Urban Develop-
   ment and the City of Ann Arbor. (Ann Arbor City Planning
   Department, Ann Arbor, MI 48108)

Missouri

Auto Club of Missouri. *What Can an Auto Club Do for Cycling?*
   St. Louis, Missouri. (Auto Club of Missouri, 3917 Lindell
   Blvd., St. Louis, MO 63108)

Montana

*University of Montana School of Forestry. 1972. *The Proposed Missoula Bikeway System.* Missoula, Montana. (School of Forestry, University of Montana, Science Complex, Missoula, MT 59801)

New Jersey

Princeton Bikeway Committee and Arnold Associates. 1973. *Princeton Bikeways: A Proposal to Implement the Bicycle Path Master Plan for Princeton, New Jersey.* Princeton, New Jersey. (Arnold Associates, 20 Nassau St., Princeton, NJ 08540)

Regional Planning Board of Princeton. 1971. *A Bicycle Path Master Plan for Princeton, New Jersey.* Princeton, New Jersey. (Regional Planning Board of Princeton, Planning Board Office, P.O. Box 390, Monument Dr., Princeton, NJ 08540)

New York

New York State Department of Transportation. 1971. "Bikeways in New York State." Albany, New York. (New York State Department of Transportation, Building 5, State Office Building Campus, Albany, NY 12226)

*________. 1972. "Erie Canal Bike & Hikeway." Albany, New York. (New York State Department of Transportation, Building 5, State Office Building Campus, Albany, NY 12226)

New York State Senate. 1973. "State of New York, 2713: A 1973-1974 Regular Session in Senate." ch. 680, Albany, New York. (New York State Department of Transportation, Building 5, State Office Building Campus, Albany, NY 12226)

*Suffolk County Planning Department. 1972. *The Suffolk County Comprehensive Bikeway Study.* Hauppauge, New York.

(Suffolk County Planning Department, Veterans Memorial
Highway, Hauppauge, NY 11787)

Ohio

*Miami Valley Regional Planning Commission. 1973. *Miami Valley
Regional Bikeway Plan.* Dayton, Ohio. (Miami Valley Regional
Planning Commission, 333 W. First St., Dayton, OH 45402)
_______ and Miami Valley Regional Bikeway Committee. 1973.
"Bikeways in the Miami Valley Region." Dayton, Ohio.
(Miami Valley Regional Planning Commission, 333 W. First
St., Dayton, OH 45402)

Oregon

*Campbell, Katharine; VanDyke, William; and Williams, Joan.
1974. *Asphalt Empire: The Oregon State Highway Division
and Transportation Planning in Oregon.* Portland, Oregon:
Oregon Student Public Interest Group (OSPIRG). (OSPIRG,
408 S.W. 2nd Ave., Portland, OR 97204)
City of Corvallis. 1973. *Bicycle Ordinance and Handbook.* Corvallis,
Oregon. (Corvallis Parks and Recreation Department, 601 S.W.
Washington Ave., Corvallis, OR 97330)
*City of Pendleton. "Pendleton Ordinances." S. 55, Pendleton,
Oregon.
Columbia Region Association of Governments. "CRAG Bikeway
Planning 1973-1974 Work Program." Portland, Oregon.
(Columbia Region Association of Governments, 6400 S.W.
Canyon Court, Portland, OR 97221)
De Leuw, Cather & Company. 1974. *Eugene Bikeway Master Plan.*
San Francisco, California. (Eugene Planning Department, City
Hall, 777 Pearl St., Eugene, OR 97401)
Eugene Bicycle Committee. "Eugene Bikeway Planning Criteria:
Being Used During 1973-1974." Eugene, Oregon. (Eugene
Bicycle Committee, 105 City Hall, 777 Pearl St., Eugene, OR
97401)

Lane Council of Governments. 1972. *Eugene-Springfield Metropolitan Area 1990 General Plan.* Eugene, Oregon. (Lane Council of Governments, 135 E. 6th St., Eugene, OR 97401)

Mid Willamette Valley Council of Governments and the Regional Parks and Recreation Agency of the Mid Willamette Valley. 1971. *Parks, Trails, Bikepaths, and the I-305 Freeway: A Linear Park for Salem, Oregon.* Salem, Oregon. (Regional Parks and Recreation Agency of the Mid Willamette Valley, 555 Liberty St., S.E., Salem, OR 97301)

*Multnomah County Division of Land Use Planning. 1974. *Citizen's Bikeway Report: East Multnomah County.* Portland, Oregon. (Multnomah County Planning Commission, 1107 S.W. Fourth Ave., Portland, OR 97204)

*Oregon Department of Transportation, Highway Division. 1973. *Oregon Bikeway Progress Report.* Salem, Oregon. (Oregon State Highway Division, Highway Building, Salem, OR 97310)

*Oregon State Highway Division. 1974. *Bikeway Design.* Salem, Oregon. (Oregon State Highway Division, Highway Building, Salem, OR 97310)

*________. *Footpaths and Bike Routes: Standards and Guidelines.* Salem, Oregon. (Oregon State Highway Division, Highway Building, Salem, OR 97310)

*Portland Bicycle Path Task Force. 1973. *Bicycle Facilities for Portland: A Comprehensive Plan.* Portland, Oregon. (Portland Planning Commission, Rm. 414, City Hall, 1220 S.W. 5th Ave., Portland, OR 97204)

*Springfield Bicycle Committee. 1974. *Springfield Bikeway Plan 1974: General Plan Phase 1.* Springfield, Oregon. (City of Springfield Department of Public Works, 346 Main St., Springfield, OR 97477)

Pennsylvania

Philadelphia Bicycle Coalition to the U.S. Environmental Protection Agency. 1973. *Facilities and Services Needed To Support Bicycle Commuting Into Center City Philadelphia.* Philadelphia: Drexel University. (U.S. Environmental Protection

Agency, Region III, 6th & Walnut Sts., Philadelphia, PA 19106)

Texas

City of Houston Department of Parks and Recreation. *Hike and Bike Story.* Houston, Texas. (City of Houston Department of Parks and Recreation, P.O. Box 1562, Houston, TX 77001)
Fort Worth Parks and Recreation Department. 1973. *Bicycle Trails: Sycamore Creek, Marine Creek, Foster/Overton.* 4th ed., Fort Worth, Texas. (Fort Worth Parks and Recreation Department, 1000 Throckmorton St., Fort Worth, TX 76102)

Washington

*City of Seattle Department of Community Development. 1972. *CompreBikeway Plan: City of Seattle.* Seattle, Washington. (City of Seattle Department of Community Development, Office of General Planning, 306 Cherry St., Seattle, WA 98104)
*Regional Planning Council and Bicycle Advocates. 1973. *Bikeways for Clark County.* Vancouver, Washington. (Clark County Regional Planning Council, 2400 T St., Vancouver, WA 98661)

Washington, D.C.

Moore, Jerry A. 1971. "Commuter Bike Paths." Washington, D.C.: Transportation Committee, District of Columbia City Council. (District of Columbia City Council, City Hall, Washington, D.C. 20004)

Wisconsin

*Bureau of Traffic Engineering and Electrical Services, D.P.W.
1971. *A Proposed System of Bicycle Commuting Routes and
Comments on the Use of Bicycles on Sidewalks.* Milwaukee,
Wisconsin. (City of Milwaukee Department of Public Works,
Bureau of Traffic Engineering and Electrical Services, Munici-
pal Building, Rm. 909, 841 No. Broadway, Milwaukee, WI
53202)
*University of Wisconsin Department of Landscape Architecture
and Environmental Awareness Center. 1969. *Milwaukee
Stadium Freeway.* Madison, Wisconsin. (Milwaukee County
Department of Public Works, Transportation Division, Court-
house Annex, Rm. 411, No. 9th St., Milwaukee, WI 53233)

Australia

*Centre for Environmental Studies, University of Melbourne. 1974.
"Integrated Park System in Greater Melbourne: A Proposed
Feasibility Study." Parkville, Victoria, Australia. (Centre for
Environmental Studies, University of Melbourne, Parkville,
Victoria, Australia)
*National Capital Development Commission: Sixteenth Annual
Report 1972-1973.* Canberra, Australia: Australian Govern-
ment Publishing Service. (National Capital Development Com-
mission, P.O. Box 373, Canberra City 2601, Australia)

Belgium

*Musée de la Vie Wallonne. 1971. *Cycles et Motocycles: Des
Origines a 1935* [Cycles and mororcycles from their origins to
1935]. Liege, Belgique. (Ministre de l'Education, Université de
Liege, L.A.S.L.A., Boulevard de la Sauveniete, 100, Liege 400,
Belgium)

Canada

*City of Calgary Engineering Department. 1972. *A Bicycle Path System for the City of Calgary.* Calgary, Alberta, Canada. (City of Calgary Transportation Planning Division, P.O. Box 2100, Calgary 2, Alberta, Canada)

*City of Calgary Transportation Department. 1971. *Commissioners' Report: Re Costs of Establishing Bicycle Rights-of-Way on City Sidewalks.* Calgary, Alberta, Canada. (City of Calgary Transportation Planning Division, P.O. Box 2100, Calgary 2, Alberta, Canada)

National Capital Commission. "The Bikeways/Les pistes cyclables." Ottawa, Canada. (Information Services, National Capital Commission, 48 Rideau, Ottawa, Ontario, K1N 8K5, Canada)

_________. 1970. "Standard Bicycle Path with Sod." Ottawa, Canada. (National Capital Commission, 48 Rideau, Ottawa, Ontario, K1L 8B9, Canada)

Germany

*Gruppe Radwegebau. 1963. *Radverkehrsanlagen Richtlinien* [Guidelines for bicycle traffic layouts]. 2d ed., Frankfurt am Rhein, Deutschland. (Gruppe Radwegebau, 5600 Wuppertal-Eberfeld, Hofaue 95/Postfach 131 471, West Germany)

Great Britain

*a'Green, George. 1953. *This Great Club of Ours: The Story of the C.T.C.* London: The Cyclists' Touring Club. (Cyclists' Touring Club, 69 Meadrow, Godalming, Surrey, Great Britain)

British Cycling Bureau, "Before the Traffic Grinds to a Halt . . ." London. (British Cycling Bureau, Greater London House, Hampstead Rd., London NW1 17QP, Great Britain)

*Claxton, Eric C. 1973. *The Elimination of Congestion, Creating Optimum Mobility and Freedom From Road Accidents in*

Existing Cities. The Hague, Netherlands: International Federation of Pedestrians. (British Cycling Bureau, Greater London House, Hampstead Rd., London NW1 17QP, Great Britain)

*―――――. *Reappraisal of Urban Mobility*. London: British Cycling Bureau. (British Cycling Bureau, Greater London House, Hampstead Rd., London NW1 17QP, Great Britain)

Cyclists' Touring Club. "Cyclist Today: A Statement of Policy." Godalming, Surrey. (Cyclists' Touring Club, 69 Meadrow, Godalming, Surrey, Great Britain)

Harlow Development Corporation. 1966. "Harlow Master Plan: Cycle and Footpath Routes." Harlow, Essex, Great Britain. (Harlow Development Corporation, Terminus House, the High, Harlow, Essex CM20 1UG, Great Britain)

*Lenthall, R. B. 1974. "Stevenage Cycleway System." Stevenage, Herts, Great Britain: Stevenage Development Corporation. (Stevenage Development Corporation, Daneshill House, Danestrete, Stevenage, Herts SG1 1XD, Great Britain)

*Peterborough Development Corporation. 1973. *Cycleways for Greater Peterborough*. Peterborough, Great Britain. (Peterborough Development Corporation, P.O. Box 3, Peterscourt, Peterborough PE1 1UJ, Great Britain)

The Netherlands

*Koninklijke Nederlandsche Toeristenbond ANWB. 1970. *Fietspaden en -oversteekplaatsen* [Bicycle paths and cyclecrossing]. La Haye, Nederland. (Koninklijke Nederlandsche Toeristenbond, Haag, La Haye, Wassenaarseweg 220, The Netherlands)

―――――. 1973. *Toeristische fietspaden een maatschapplijk belang* [Tourist bikepaths, a social interest]. La Haye, Nederland. (Koninklijke Nederlandsche Toeristenbond, Haag, La Haye, Wassenaarseweg 220, The Netherlands)

*Stichting Recreatie. 1974. *Fiets en recreatie* [The bicycle and recreation]. La Haye, Nederland. ("stichting: fiets," secretariaat euopaplein 2, Amsterdan-z, The Netherlands)

*"stichting: fiets." *Meer fietsen . . . meer fietspaden* [More bicycles—
    more bicycle paths]. Amsterdan, Nederland. ("stichting:
    fiets," secretariaat euopaplein 2, Amsterdan-z, The
    Netherlands)

*________. 1972. *Optimal aandacht voor fietsers* [Optimal consider-
    ation for bicyclists]. Amsterdan, Nederland. ("stichting:
    fiets," secretariaat euopaplein 2, Amsterdan-z, The
    Netherlands)

*________. 1973. *Voorzieningen voor het parkeren en onderbregen
    van fietsen -overzicht van humpmiddelen, fuctionele en
    technische oplossingen* [Provision for the parking and shelter
    of bicycles]. Amsterdan, Nederland. ("stichting: fiets,"
    secretariaat euopaplein 2, Amsterdan-z, The Netherlands)

Sweden

*Rosen, Nils. 1973. "Planning Pedal Power: The Bicycle in the
    City." New York: Swedish Information Service. (Swedish
    Information Service, 1960 Jackson St., San Francisco, CA
    94109)

*Stenman, Bo. 1972. "Traffic Environment in Swedish Urban
    Areas." Stockholm: Royal Ministry for Foreign Affairs,
    Swedish Information Service. (Swedish Information Service,
    1960 Jackson St., San Francisco, CA 94109)

BOOKS

Botanical References

Chauvin, Rémy, 1967. *The World of an Insect.* Translated by
    Harold Oldroyd. New York: McGraw-Hill Book Co.

Davis, Donald D. 1973. *Air Pollution Damages Trees.* Washington,
    D.C.: U.S. Department of Agriculture.

Hindawi, Ibrahim Joseph. 1970. *Air Pollution Injury to Vegetation.*
    Washington, D.C.: U.S. Department of Health, Education, and
    Welfare (No. AP-71).
Hottes, Alfred C. 1942. *The Book of Trees.* New York: A. T. De La
    Mare Co.
Robinette, Gary O. 1972. *Plants/People/and Environmental
    Quality: A Study of Plants and Their Environmental
    Functions.* Washington, D.C.: U.S. Department of the Interior,
    National Park Service, and American Society of Landscape
    Architects Foundation.
*Transit Planting: A Manual for the Horticultural Development of
    Public Transportation Environments.* 1974. Washington, D.C.:
    The American Horticultural Society, Urban Mass Transporta-
    tion Administration, and U.S. Department of Transportation.
    Stock No. VA-06-0006.
Williamson, Joseph F., ed. 1971. *Sunset Western Garden Book.*
    Menlo Park, California: Lane Magazine and Book Co.
Wyman, Donald. 1971. *Wyman's Gardening Encyclopedia.* New
    York: Macmillan Co.

Historical References

Hoyt, Hugh Myron, Jr. 1966. *The Good Road Movement in Oregon
    1900-1920.* Ph.D. dissertation, University of Oregon.
Leonard, Irving A. 1969. *When Bikehood Was in Flower: Sketches
    of Early Cycling.* South Tamworth, New Hampshire: Bear-
    camp Press.
Mason, Philip P. 1957. *The League of American Wheelmen and the
    Good-Roads Movement 1880-1905.* Ph.D. dissertation, Uni-
    versity of Michigan.
Smith, Robert A. 1972. *A Social History of the Bicycle: Its Early
    Life and Times in America.* New York: American Heritage
    Press.
St. Pierre, Roger. 1973. *The Book of the Bicycle.* London: Triune
    Books.

General References

Brooks, Mary E. 1969. *Planning for Urban Trails.* Chicago: American Society of Planning Officials, Planning Advisory Service.

City of Eugene. 1973. *Neighborhood Organization Policy: For Citizen Planning Group.* Eugene, Oregon: Eugene Comprehensive Planning Department.

Drapela, Ernest, and Pratt, Kevin. 1972. *30 Bike Rides in Lane County.* Eugene, Oregon: Velodrome Books.

A Handbook on Bicycle Tracks and Cycle Racing. Dayton, Ohio: Huffman Manufacturing Co.

Jankowski, Nick, and Jankowski, Elske. 1973. *55 Oregon Bicycle Trips.* Beaverton, Oregon: Touchstone Press.

Kleinsasser, William. 1974. *Experiential Design Considerations.* Eugene, Oregon: University of Oregon. Unpublished.

Patrick, John, ed. 1971. *Pedaler's Pamphlet: A Collection of Things About Bicycles for the People of Eugene.* Eugene, Oregon: Canterbury Press.

Ramsdell, Charles. 1968. *Special Supplement to the HemisFair Edition of San Antonio: A Historical and Pictorial Guide.* Austin, Texas: University of Texas Press.

Tunnard, Christopher, and Reed, Henry Hope. 1956. *American Skyline: The Growth and Form of Our Cities and Towns.* New York: Mentor Books.

Parks and Recreation References

Cook, Walter L. 1965. *Bike Trails and Facilities: A Guide to Their Design, Construction, and Operation.* Wheeling, West Virginia: American Institute of Park Executives.

U.S. Department of Interior, Bureau of Outdoor Recreation. 1970. *Federal Assistance in Outdoor Recreation.* Washington, D.C.: U.S. Government Printing Office. Stock No. 398-495.

U.S. Department of Interior, Federal Highway Administration. *Proceedings: National Symposium on Trails, Washington, D.C., June 2-6.* Washington, D.C.: U.S. Government Printing Office. Stock No. 2416-0042.

U.S. Department of Interior and U.S. Department of Transportation. 1972. *Bicycle for Recreation and Commuting.* Washington, D.C.: U.S. Government Printing Office. Stock No. 461-880.

Transportation References

California State Senate. 1972. *State of California Vehicle Code 1972.* Sacramento, California: Department of Motor Vehicles.

Kennedy, Norman; Kell, James M.; and Homburger, Wolfgang S. 1962. *Fundamentals of Traffic Engineering.* Berkeley, California: Institute of Transportation and Traffic Engineering, University of California. UC Syllabus Series No. 404.

Leinwand, Gerald. 1969. *Problems of American Society: The Traffic Jam.* New York: Washington Square Press.

Oregon State Senate. 1973. *Oregon Revised Statutes.* vol. 3, ch. 326-495. Legislative Counsel Committee.

Stone, Tabor R. 1971. *Beyond the Automobile: Reshaping the Transportation Environment.* Englewood Cliffs, New Jersey: Prentice-Hall, Inc.

Wilbur Smith and Associates. 1970. *The Potential for Bus Rapid Transit.* Detroit: Automobile Manufacturers Association.

U.S. Department of Transportation, Federal Highway Administration. 1972. *Manual of Uniform Traffic Control Devices for Streets and Highways.* Washington, D.C.: U.S. Government Printing Office. Stock No. 5001-0021.

PERIODICALS AND PAMPHLETS

Botanical References

Deering, Robert B., and Brooks, Frederick A. 1954. "The Effect of Plant Material Upon the Microclimate of House and Garden," In *The National Horticultural Magazine,* July, pp. 162-67.

Everett, Michael D. 1974. "Roadside Air Pollution Hazards in Recreational Land Use Planning." In *Journal of the American Institute of Planners,* March, pp. 83-88.

Shurleff, Malcolm C. 1974. "How Air Pollution Affects Tree Varieties." In *Grounds Maintenance: The Technical Magazine of Landscape Design, Construction, and Maintenance.* September, pp. 20-24, 49-50.

Historical References

Good Roads and Cyclist: Organ of the Oregon Road Club and League of American Wheelmen. May, 1896-April, 1897.

Harmon, Phyllis. 1974. "L.A.W.: On the Road Since 1880." In *Bicycle Spokesman Magazine,* March, p. 48.

General References

Amateur Bicycle League of America. 1973-1974. *Cyclenews: A Journal of Bicycle Racing.* vol. 2 (6)-vol. 3 (2).

"La bicyclette, nouvelle toquade des Américains" [The bicycle, new American infatuation]. In *Information du Commerce Exterieur,* October 25, 1972, pp. 3-5.

Canadian Cycling Association. 1974. *Canadian Cyclist-Cyclist Canadien.* Vanier City, Ontario, January-February.

Cyclists' Touring Club. 1973-1974. *Cycletouring: Magazine of the C.T.C.* Godalming, Surrey, Great Britain, December-January.

DeLong, A. Fred. "Deraileur Lightweights: A New Dimension in
     Cycling." Chicago: Schwinn Bicycle Co.
Desimone, Vincent R. 1972. "Planning Criteria for Bikeways." In
     Transportation Engineer. Milwaukee, Wisconsin: Automobile
     Club of Southern California, American Society of Civil Engi-
     neers, National Transportation Engineering Meeting, pp. 1-25.
"En 1972, les Américains ont acheté pour la premiere fois plus de
     bicyclettes que de voitures" [In 1972, Americans bought more
     bicycles than cars]. 1973. In *Information du Commerce
     Exterieur,* October 17, pp. 4-5.
Forester, John. 1973. "What About Bikeways?" In *Bike World,*
     February, p. 36.
Friends for Bikecology. 1971. "Bikecology Overview." Santa
     Barbara, California, pp. 1-11.
———. 1973. "The Bikeway Movement: A Sad State of Affairs."
     In *Serendipity* 9, summer.
Hanneman, Ralph. 1973. "Bicycle Commuting: The Push Is On by
     Urban Environmental Activists." New York: Bicycle Institute
     of America.———
"Here Come the Bikeways: If Your Community Has Not Yet Faced
     the Realities of the Adult Bike, This Report May Be Helpful."
     1972. In *Sunset: Magazine of Western Living,* October, pp. 94-
     97.
International City Management Association. 1973. *Management
     Information Service Report: Planning and Development of
     Bikeway Systems,* vol. 5, April, Washington, D.C.
Nelson, Thomas. 1972. "To Protect a Shoreline." In *Landscape
     Architecture,* October, pp. 51-52.
"What About Bike Tripping?—A Joy, a Torture, and Certainly an
     Experience You Won't Forget." 1974. In *Sunset: Magazine of
     Western Living,* May, pp. 79-85.
White, Paul Dudley. 1964. "We Need More Bicycle Paths . . . Let's
     Go!" Chicago: Schwinn Bicycle Co.

Legal References

Bicycle Institute of America. 1973. "California Passes Bikeway
     Funding Law." In *Boom in Bikeways,* October.

________. 1973. "State Bicycle Legislation Proliferating." In *Boom in Bikeways,* October.

________. "Summary of 1973 Bikeway Legislation: Will Cyclists in Your State Benefit?" New York.

Franz, Clifford L., and DeLong, A. Fred. 1974. "California Traffic Safety Education To Include All Road Users." In *L.A.W. Bulletin* (League of American Wheelmen), January, pp. 7-8.

Friends for Bikecology. 1972. "Legislation." In *Serendipity,* winter.

________. 1973. "Startling Statements." In *Serendipity,* fall.

________. 1974. "An Anti-Bikeway Movement." In *Serendipity,* winter.

Stevenson, M. Lynn. 1973. Presentation at Clear Air Act Public Hearing, San Francisco, May 21.

"What State Are You In?" In *Bicycle Spokesman Magazine,* March, 1974, p. 48.

HOUSE BILL 1700 (OREGON)

STATE LEGISLATION
AN ACT
Relating to ways for public travel; creating new provisions; and
amending ORS 366.515, 366.525 and 366.790.

BE IT ENACTED BY THE PEOPLE OF THE STATE OF
OREGON:

SECTION 1. Section 2 of this Act is added to and made a part
of ORS chapter 366.

SECTION 2. (1) Out of the funds received by the commission
or by any county or city from the State Highway Fund reasonable
amounts shall be expended as necessary for the establishment of
footpaths and bicycle trails. Footpaths and bicycle trails shall be
established wherever a highway, road or street is being constructed,
reconstructed or relocated. Funds received from the State Highway
Fund may also be expended to maintain such footpaths and trails
and to establish footpaths and trails along other highways, roads
and streets and in parks and recreation areas.

(2) Footpaths and trails are not required to be established
under subsection (1) of this section:

(a) Where the establishment of such paths and trails would be
contrary to public safety;

(b) If the cost of establishing such paths and trails would be
excessively disproportionate to the need or probable use; or

(c) Where sparsity of population, other available ways or other
factors indicate an absence of any need for such paths and trails.

(3) The amount expended by the commission or by a city or
county as required or permitted by this section shall never in any
one fiscal year be less than one percent of the total amount of the
funds received from the highway fund. However:

(a) This subsection does not apply to a city in any year in which the one percent equals $250 or less, or to a county in any year in which the one percent equals $1,500 or less.

(b) A city or county in lieu of expending the funds each year may credit the funds to a financial reserve or special fund in accordance with ORS 280.100, to be held for not more than 10 years, and to be expended for the purposes required or permitted by this section.

(4) For the purposes of this chapter, the establishment of paths and trails and the expenditure of funds as authorized by this section are for highway, road and street purposes. The commission shall, when requested, provide technical assistance and advice to cities and counties in carrying out the purpose of this section. The division shall recommend construction standards for footpaths and bicycle trails. The division shall, in the manner prescribed for marking highways under ORS 483.040, provide a uniform system of signing footpaths and bicycle trails which shall apply to paths and trails under the jurisdiction of the commission and cities and counties. The commission and cities and counties may restrict the use of footpaths and bicycle trails under their respective jurisdiction to pedestrians and nonmotorized vehicles.

(5) As used in this section, "bicycle trail" means a publicly owned an maintained lane or way designated and signed for use as a bicycle route.

SECTION 3. ORS 366.515 is amended to read:

366.515. (1) The highway fund shall be expended under the jurisdiction of the commission.

(2) Except as provided in ORS 367.236 and 366.735, the commission shall set aside from the highway fund, in the following order:

(a) An amount sufficient for the salaries and expenses of the highway department.

(b) A sufficient amount to cover the cost of operating and maintaining state highways which have been constructed or improved.

(c) Sufficient funds to meet the Federal Government appropriation entitled "An Act to provide that the United States shall aid the States in the construction of rural post roads and for other purposes," or any federal appropriation that may be provided.

(d) The remainder shall be used for any of the purposes authorized by law.

(3) All the highway fund not otherwise specifically applied shall be expended by the commission in its discretion, EXCEPT AS REQUIRED BY SECTION 2 OF THIS 1971 ACT, on the construction, maintenance, betterment or pavement of roads and highways within the state.

SECTION 4. ORS 366.525 is amended to read:

366.525. There shall be and hereby are appropriated out of the highway fund annually such sums of money as will equal 20 percent of all moneys credited to the State Highway Fund by the State Treasurer between July 1 of any year and June 30 of the following year and which have accrued from funds transferred to the highway fund by the State Treasurer under ORS 481.950, paragraph (b) of subsection (2) of ORS 484.250 and ORS 767.635. The appropriation shall be distributed among the several counties for the purposes (now) provided by law.

SECTION 5. ORS 366.790 is amended to read:

366.790. Money paid to cities under ORS 366.785 to 366.820 shall be used only for the purposes stated in section 3, Article IX of the Oregon Constitution AND STATUTES ENACTED PURSUANT THERETO INCLUDING SECTION 2 OF THIS 1971 ACT.

SECTION 55, PENDLETON ORDINANCES (OREGON)

SECTION 55. All bicycle operators under the age of 18 years shall take a written examination as prescribed by The City of Pendleton and obtain a passing grade of 80 percent, no registration certificate may be obtained until the provisions of this ordinance are complied with.

SENATE BILL #821, CALIFORNIA

PUBLIC UTILITIES–TRANSPORTATION DEVELOPMENT
CHAPTER 740
Senate Bill No. 821

An act to amend Sections 99232, 99233, 99241, 99267, and 99401, and to add Section 99234 to the Public Utilities Code relating to transportation.

LEGISLATIVE COUNSEL'S DIGEST

Requires up to 2 percent of the funds, after allocations for specified administrative costs and transportation planning, allocated for transportation purposes under the Mills-Alquist-Deddeh Act to be made available to cities and counties for facilities provided for the exclusive use by pedestrians and bicycles prior to making such funds available for public transportation purposes and local street and road purposes, unless the designated transportation agency makes a specific finding.

Authorizes the use of the above allocations for pedestrian and bicycle facilities in all counties (including counties with a population of 500,000 or more and cities within such counties in areas of such counties served by public transportation systems).

Specifies that such allocations for an area may be applied by an operator serving that area to satisfy the requirement that 75 percent of the funds allocated to the operator each fiscal year under that act be used for capital expenditures.

Authorizes any local entity contributing money to an operator to designate that portion of its annual contributions but not to exceed one-half thereof, may be applied by the operator to satisfy the above 75 percent requirement.

Makes related changes.

Makes additional changes in Sec. 99267, Public Utilities Code, proposed by SB 314, to be operative only if this bill and SB 314 are both chaptered and effective January 1, 1974, and this bill is chaptered after SB 314.

The people of the State of California do enact as follows:

SECTION 1. Section 99232 of the Public Utilities Code is amended to read:
99232.

For counties with a population of 500,000 or more, as determined under Section 28020 of the Government Code, as now or hereafter amended but excluding counties with more than 4,500 miles of maintained county roads, the amount representing the apportionments of the areas of all operators shall be available solely for claims for Section 99234 purposes and for Article 4 (commencing with Section 99260) purposes and any such moneys not allocated in any year shall be available for such claims in subsequent years.

SECTION 2. Section 99233 of the Public Utilities Code is amended to read:
99233.

The fund shall be allocated by the designated transportation planning agency in accordance with the following priorities:

(a) First, there shall be allocated to the county such sums as are necessary for the county to administer this chapter.

(b) Thereafter there shall be allocated to the transportation planning agency such sums as are necessary to administer this chapter.

(c) Thereafter there shall be allocated to the transportation planning agency if it is statutorily created or is serving as the California representation to an agency created by interstate compact, such sums as it may approve up to 3 percent of annual revenues for the conduct of the transportation planning process, unless a greater amount is approved by the secretary.

(d) Thereafter there shall be allocated to cities and counties such moneys as are approved by the transportation planning agency for claims presented pursuant to Section 99234.

(e) Thereafter there shall be allocated to operators such moneys as are approved by the transportation planning agency for claims presented pursuant to Article 4 (commencing with section 99260) of this chapter.

(f) Thereafter there shall be allocated to cities and counties such moneys as are approved by the transportation planning agency for claims presented pursuant to Article 8 (commencing with section 99400) of this chapter.

SECTION 3. Section 99234 is added to the Public Utilities Code, to read:
99234.

After allocations for the purposes specified in subdivisions (a), (b), and (c) of Section 99233, 2 percent of the remaining moneys shall be made available to cities and counties for facilities provided for the exclusive use by pedestrians and bicycles, unless the transportation planning agency finds that such moneys could be used to better advantage for the purposes stated in Article 4 (commencing with Section 99260) of this chapter, or for local street and road purposes in those areas where such moneys may be expended for such purposes, in the development of a balanced transportation system.

Claims for such facilities shall be filed in the same manner as are claims filed for the purposes of subdivision (a) of Section 99400.

SECTION 4. Section 99241 of the Public Utilities Code is amended to read:
99241.

Except for allocations made for purposes of Section 99234 and subdivision (a) of Section 99400, which shall be subject to the rules and regulations adopted by the transportation planning agency, all matters necessary and convenient to the implementation of this chapter shall be subject to rules and regulations, consistent with statute, adopted by the secretary, with the advice and consent of the State Transportation Board, and those rules and regulations may be revised from time to time.

Such rules and regulations shall specify the procedures by which evaluation and review by the transportation planning agency of public transportation claims shall be accomplished, and shall require submission of corresponding budgets or financial plans, certified financial statements, and other information required in connection therewith. The rules and regulations shall provide for the orderly and periodic distribution of moneys in the fund so that the areas served by the operator will be provided public transportation services on a continuing basis and so that there will be an orderly improvement and maintenance of the system of the operator. The rules and regulations shall provide for the approval of sufficient moneys from the local fund to accomplish the intent of the Legislature as expressed in the findings and declarations in Section 99220.

Such rules and regulations may require that the transportation planning agency, in reviewing claims, give due consideration to the level of the operator's passenger fares and charges, the efficiency of the operator's operations and operating policies and practices, the extent to which the operator is meeting the transportation needs of the area served, and the extent to which the operator is making full use of other available revenues and funds, including federal transportation grants.

SECTION 5. Section 99267 of the Public Utilities Code is amended to read:
99267.

(a) Except for a transit district or a municipal operator during its first five years of existence, if it did not operate a public transportation system prior to July 1, 1972, at least 75 percent of funds received under this article each fiscal year shall be used for capital expenditures. Such capital expenditures shall consist of acquisition of land and other real property, current acquisition or replacement of transportation vehicles or conveyances, and acquisition, construction, enlargement, or repair of property and facilities incidental to or necessary or convenient in connection with the foregoing, depreciation, and payment of principal and interest on its bonded indebtedness, equipment trust certificates, or other indebtedness, including any amounts in the accomplishment of a defeasance under any outstanding revenue bond indenture.

(b) The amount of federal and other state funds granted or approved for capital expenditures on a matching basis by the operator, the local contributions designated pursuant to subdivision (c) and the allocations made pursuant to Section 99234 for the area served by the operator in any fiscal year may be applied by the operator to satisfy the requirement of subdivision (a).

(c) Any local entity contributing money to an operator, during the 1973-74 fiscal year or any subsequent fiscal year, may designate what portion of its annual contribution, but not to exceed one-half thereof, may be applied to satisfy the requirement specified in subdivision (a), whether or not such portion is to be expended for capital expenditures.

(d) When unique and unusual circumstances arise whereby funds are available for capital expenditures but not for other more urgent costs, an operator may file an application with the secretary for a temporary waiver which, if granted, shall authorize funds otherwise restricted for capital expenditures to be used for other more urgent purposes. To approve such a request, the secretary shall determine that financing the noncapital expenditures by other means, such as fare increases or curtailment of services, will adversely affect public transportation service for the area.

SECTION 5.5. Section 99267 of the Public Utilities Code is amended to read:
99267.

(a) Except for a transit district or a municipal operator during its first five fiscal years in the operation of a public transportation system, if it did not operate a public transportation system prior to July 1, 1972, at least 75 percent of funds received under this article shall be used for capital expenditures. Such capital expenditures shall consist of acquisition of land and other real property, current acquisition or replacement of transportation vehicles, or conveyances, and acquisition, construction, enlargement, or repair of property and facilities incidental to or necessary or convenient in connection with the foregoing, depreciation, and payment of principal and interest on its bonded indebtedness, equipment trust certificates, or other indebtedness, including any amounts in the accomplishment of a defeasance under any outstanding revenue bond indenture.

(b) The amount of federal and other state funds granted or approved for capital expenditures on a matching basis by the operator, the local contributions designated pursuant to subdivision (c), and the allocations made pursuant to Section 99234 for the area served by the operator in any fiscal year may be applied by the operator to satisfy the requirement of subdivision (a).

(c) Any local entity contributing money to an operator, during the 1973-74 fiscal year or any subsequent fiscal year, may designate what portion of its annual contribution, but not to exceed one-half thereof, may be applied to satisfy the requirement specified in subdivision (a), whether or not such portion is to be expended for capital expenditures.

(d) When unique and unusual circumstances arise whereby funds are available for capital expenditures but not for other more urgent costs, an operator may file an application with the secretary for a temporary waiver which, if granted, shall authorize funds otherwise restricted for capital expenditures to be used for other more urgent purposes. To approve such a request, the secretary shall determine that financing the noncapital expenditures by other means, such as fare increases or curtailment of services, will adversely affect public transportation service for the area.

SECTION 6. Section 99401 of the Public Utilities Code is amended to read:

99401.

The transportation planning agency shall adopt rules and regulations delineating procedures for the submission of claims under Section 99234 and subdivision (a) of Section 99400 and stating criteria by which they will be analyzed and evaluated. Such rules and regulations shall provide for orderly and periodic distributions of moneys.

To the extent necessary to perform its duties under this article, the transportation planning agency shall have full access to the books, records, and accounts of the claimant cities and counties.

SECTION 7. It is the intent of the Legislature, if this bill and Senate Bill No. 314 are both chaptered and become effective January 1, 1974, both bills amend Section 99267 of the Public Utilities Code, and this bill is chaptered after Senate Bill No. 314, that the amendments to Section 99267 proposed by both bills be given effect and incorporated in Section 99267 in the form set forth in Section 5.5 of this act. Therefore, Section 5.5 of this act shall become operative only if this bill and Senate Bill No. 314 are both chaptered and become effective January 1, 1974, both amend Section 99267, and this bill is chaptered after Senate Bill No. 314, in which case Section 5 of this act shall not become operative.

Approved and filed Sept. 25, 1973.

SENATE BILL #36, CALIFORNIA

STATE HIGHWAYS—BICYCLE FACILITIES
CHAPTER 1092
Senate Bill No. 36

An act to amend Section 2106 and 2106.5 of, to add Section 100.13 and 149.2 to, and to add Chapter 8 (commencing with Section 2370) to Division 3 of the Streets and Highways Code relating to transportation.

LEGISLATIVE COUNSEL'S DIGEST

Requires the California Highway Commission and the Department of Public Works to set aside in each of its annual budget reports an amount of not less than $360,000 for the construction of bicycle facilities to be used in conjunction with the state highway system.

Requires the department to prepare and submit an annual report to the Legislature summarizing programs it has undertaken for the development of bicycle lanes and exclusive busways, together with information re availability of federal funds for, and fiscal impact of federal participation in, such programs.

Creates a "Bicycle Lane Account" in the State Transportation Fund, and requires that $30,000 be transferred monthly to such account from the net revenues derived from 1.04 cents per gallon tax imposed under the Motor Vehicle Fuel License Tax Law allocated to the cities and the counties.

Continuously appropriates the money in the "Bicycle Lane Account" to the department for allocation for city and county bicycle lane projects in accordance with a prescribed schedule of priorities and subject to specified limitations, and for administrative costs of the department in connection with such allocations.

Requires the department to adopt necessary rules and regulations to implement the above provisions re city and county bicycle lane projects.

Makes additional changes in Section 2106, Streets and Highways Code, proposed by AB 945, to be operative only if AB 945 and this bill are both chaptered, and this bill is chaptered after AB 945.

The people of the State of California do enact as follows:

SECTION 1. Section 100.13 is added to the Streets and Highways Code, to read:

100.13

In each annual budget report prepared by the commission and the department under Section 143.1, an amount of not less than three hundred sixty thousand dollars ($360,000) shall be set aside for the construction of bicycle facilities to be used in conjunction with the state highway system.

SECTION 2. Section 149.2 is added to the Streets and Highways Code, to read:

149.2.

Prior to March 1 of each year, the department shall prepare and submit an annual report to the Legislature summarizing programs it has undertaken for the development of bicycle lanes and exclusive busways. Such report shall also summarize the existing directive received by the department from the Federal Highway Administration concerning the availability of federal funds for such programs, together with an estimate of the fiscal impact of such federal participation in the programs.

SECTION 3. Section 2106 of the Streets and Highways Code is amended to read:

2106.

A sum equal to the net revenue derived from one and four one-hundredths cent ($0.0104) per gallon tax under the Motor Vehicle Fuel License Tax Law shall be apportioned monthly from the Highway Users Tax Fund among counties and cities for use exclusively for county road and city street purposes, as provided in this section.

The amounts available under this section shall be apportioned, as follows:

(a) Four hundred dollars ($400) per month shall be apportioned to each city and city and county and eight hundred dollars ($800) per month shall be apportioned to each county and city and county.

(b) Thirty thousand dollars ($30,000) per month shall be transferred to the Bicycle Lane Account in the State Transportation Fund.

(c) The balance shall be apportioned as follows:

(1) A base sum shall be computed for each county by using the same proportions of fee-paid and exempt vehicles as are established for purposes of apportionment of funds under subdivision (d) of Section 2104.

(2) For each county the percentage of the total assessed valuation of tangible property subject to local tax levies within the county which is represented by the assessed valuation of tangible property outside the incorporated cities of the county shall be applied to its base sum, and the resulting amount shall be apportioned to the county. The assessed valuation of taxable tangible property for purposes of this computation shall be that most recently used for countywide tax levies as reported to the State Controller by the State Board of Equalization. In the event an incorporation or annexation is legally completed following the base sum computation, the new city's assessed valuation shall be deducted from the county's assessed valuation the estimate of which shall be provided by the State Board of Equalization.

(3) The difference between the base sum for each county and the amount apportioned to the county shall be apportioned to the cities of that county in the proportion that the population of each city bears to the total population of all cities in the country. Populations used for determining expenditure of moneys under Section 2107 are to be used for purposes of this section.

No apportionment shall be made to any city or county which does not have a select system established in accordance with the provisions of Section 186.3. If all or a portion of the select system of a city or county shall be a change in the physical limits of such city or county be included within the boundaries of another city or county, such portion shall be considered to be a part of the select system of such other city or county. Any amounts which would otherwise have been apportioned to a city which does not so qualify for an apportionment shall be reapportioned to the remaining cities in the county in which such city is located, or, if there are no eligible cities in the county, to the county. Any amounts which would otherwise have been apportioned to a county which does not so qualify for an apportionment shall be reapportioned to the eligible cities in such county, or, if there are no eligible cities in the county, to other counties and cities pursuant to this section.

SECTION 4. Section 2106.5 of the Streets and Highways Code is amended to read:

2106.5.

(a) Each county and any of its incorporated cities may enter into an agreement regarding the base sum established by paragraph (1) of subdivision (s) of Section 2106, providing for expenditure of the amounts apportioned to the county and apportioned for expenditure within the cities participating in the agreement upon selected roads and streets within the county and the cities participating in the agreement included in the select system of county roads and city streets for such county and cities established as provided in Section 186.3.

(b) Any of the incorporated cities within a county may enter into an agreement among themselves regarding the amount apportioned to them pursuant to paragraph (3) of subdivision (c) of Section 2106 for expenditure upon selected streets included in the select system of city streets, established as provided in Section 186.3, within the cities participating in the agreement.

(c) Any such agreement shall be filed with the Controller. After verification of the agreement by the Controller, the Controller shall make disposition of the apportionments to the parties participating in the agreement in accordance with terms of the agreement.

SECTION 5. Chapter 8 (commencing with Section 2370) is added to Division 3 of the Streets and Highways Code, to read:

CHAPTER 8. BICYCLE LANES

2370

It is the intent of the Legislature in enacting this chapter to provide a continuing source of state funding to aid cities and counties in obtaining funds from federal sources for the construction of, and the acquisition of rights-of-way for, bicycle lanes, where the separation of bicycle traffic from motor vehicle traffic would increase the traffic capacity or safety of a highway.

2371.

The Bicycle Lane Account is hereby created in the State Transportation Fund. The money in the Bicycle Lane Account is continuously appropriated to the department for carrying out the purposes of this chapter.

2372.

To the extent funds are available, the department shall allocate from the Bicycle Lane Account funds for the following purposes and in accordance with the following priorities:

(a) To the department such sums as are necessary to administer this chapter.

(b) To cities and counties for each eligible bicycle lane project, as determined by the department, an amount equal to one-half of the estimated amount required as local contribution in order to obtain available federal funds for the construction of, including the acquisition of necessary rights-of-way for such project. The amount allocated pursuant to this subdivision shall be increased if an increase is necessary to reduce the city or country contribution to the project to one-third of the estimated cost of the project.

(c) To the cities and counties for each nonfederally funded eligible bicycle lane project, as determined by the department, an amount equal to two-thirds of the estimated cost of construction of, including the acquisition of necessary rights-of-way for the project.

2373.

(a) The maximum amount the department may allocate to an applicant city or county with an eligible project during any particular fiscal year shall be an amount equal to the amount available for allocation during that fiscal year times the ratio of the applicant's population to the total population of all the cities and counties with eligible projects. For purposes of this subdivision, the population of a county shall be the population of its unincorporated area.

(b) Funds in excess of the amount determined pursuant to subdivision (a) may be allocated by the department to a city or county with an eligible project, if funds are available after the requests of all cities and counties with eligible projects are met.

2374.

If funds are insufficient to completely finance any eligible project, such a project shall be deemed an eligible project for allocations in subsequent fiscal years.

2375.

The department shall adopt necessary rules and regulations to implement the provisions of this chapter. Such rules and regulations shall contain provisions for establishing priorities in the event that

the amount of funding needed for eligible projects exceeds the
amount of available funds. In preparing such rules and regulations
and in administering this chapter, the department shall periodically
confer with the Director of Parks and Recreation and with the committee established pursuant to Section 2315.

SECTION 6, Section 2106 of the Streets and Highways Code
is amended to read:
2106.

A sum equal to the net revenue derived from one and four
one-hundredths cent ($0.0104) per gallon tax under the Motor
Vehicle Fuel License Tax Law shall be apportioned monthly from
the Highway Users Tax Fund among counties and cities for use
exclusively for county road and city street purposes, as provided in
this section.

The amounts available under this section shall be apportioned
as follows:

(a) Four hundred dollars ($400) per month shall be appor-
tioned to each city and city and county and eight hundred dollars
($800) per month shall be apportioned to each county and city and
county.

(b) Thirty thousand dollars ($30,000) per month shall be
transferred to the Bicycle Lane Account in the State Transportation
Fund.

(c) The balance shall be apportioned as follows:

(1) A base sum shall be computed for each county by using
the same proportions of fee-paid and exempt vehicles as are established for purposes of apportionment of funds under subdivision
(d) of Section 2104.

(2) For each county the percentage of the total assessed valuation of tangible property subject to local tax levies within the
county which is represented by the assessed valuation of tangible
property outside the incorporated cities of the county shall be applied to its base sum, and the resulting amount shall be apportioned
to the county. The assessed valuation of taxable tangible property
for purposes of this computation shall be that most recently used
for countywide tax levies as reported to the State Controller by the
State Board of Equalization. In the event an incorporation or
annexation is legally completed following the base sum computation, the new city's assessed valuation shall be deducted from the

county's assessed valuation, the estimate of which may be provided by the State Board of Equalization.

(3) The difference between the base sum for each county and the amount apportioned to the county shall be apportioned to the cities of that county in the proportion that the population of each city bears to the total population of all cities in the county. Populations used for determining expenditure of moneys under Section 2107 are to be used for purposes of this section.

No apportionment shall be made to any city or county which does not have a select system established in accordance with the provisions of Section 186.3. If all or a portion of the select system of a city or county shall be a change in the physical limits of such city or county be included within the boundaries of another city or county, such portion shall be considered to be a part of the select system of such other city or county. Any amounts which would otherwise have been apportioned to a city which does not so qualify for an apportionment shall be reapportioned to the remaining cities in the county in which such city is located, or if there are no eligible cities in the county, to the county. Any amounts which would otherwise have been apportioned to a county which does not so qualify for an apportionment shall be reapportioned to the eligible cities in such county, or if there are no eligible cities in the county, to other counties and cities pursuant to this section.

SECTION 7. It is the intent of the Legislature, if this bill and Assembly Bill No. 945 are both chaptered and amend Section 2106 of the Streets and Highways Code, and this bill is chaptered after Assembly Bill No. 945, that the amendments to Section 2106 proposed by both bills be given effect and incorporated in Section 2106 in the form set forth in Section 6 of this act. Therefore, Section 6 of this act shall become operative only if this bill and Assembly Bill No. 945 are both chaptered, both amend Section 2106, and Assembly Bill No. 945 is chaptered before this bill, in which case Section 3 of this act shall not become operative.

Approved and filed August 22, 1972.

VEHICLE CODE, STATE OF CALIFORNIA

Section 39003, Disposition of License Fees

39003. Funds derived from bicycle license fees shall be retained by the licensing city or county and used to implement and improve bicycle registration and safety programs in its jurisdiction, except that revenues which, prior to the effective date of this division, were used for other purposes may continue to be used for the same purposes.

Added Ch. 885, Stats. 1972, Effective Aug. 15, 1972, by terms of an urgency clause printed following Sec. 39000.

Bruce L. Balshone is Director of the Oregon Bicycle Transit Study Committee within the Departments of Urban Planning and Landscape Architecture at the University of Oregon. Paul L. Deering and Brian D. McCarl are currently associated with the Committee. Mr. Balshone has served as an advisor to many local governmental planning agencies within Oregon and Washington in helping to formulate bicycle planning criteria in their local communities.

Mr. Balshone has traveled extensively within Northern California, Oregon, and Washington as well as Western Europe touring and observing many bicycle systems. Mr Deering has resided for much of his life in Davis, California, where he learned to enjoy the city's simple and complete bicycle network.

Mr. Balshone has received a bachelor's degree in Sociology and a master's degree in Landscape Architecture from the University of Oregon. Mr. Deering currently holds a bachelor's degree in Landscape Architecture from the University of Oregon, while Mr. McCarl is presently enrolled in the undergraduate program in Landscape Architecture at the University of Oregon.

AIR QUALITY MANAGEMENT AND LAND USE
PLANNING: Legal, Administrative, and Methodological
Perspectives
George Hagevik, Daniel R. Mandelker
and Richard K. Brail

DEVELOPMENT ON A HUMAN SCALE: Potentials for
Ecologically Guided Growth in Northern New Mexico
Peter van Dresser

SPATIAL DESIGN AND PLANNING IN THE UK: Its
Relevance to Developing Countries
edited by Robert J. Marshall

STATE ENVIRONMENTAL MANAGEMENT: Case Studies
of Nine States
Elizabeth H. Haskell
and Victoria S. Price